T0208621

Lecture Notes in Mathematics 2174

More information about this series at http://www.springer.com/series/304

W.A. Zúñiga-Galindo

Pseudodifferential Equations Over Non-Archimedean Spaces

 Springer

W.A. Zúñiga-Galindo
Department of Mathematics
Center for Research and Advanced Studies
    of the National Polytechnic Institute
    (CINVESTAV)
Mexico City, Mexico

ISSN 0075-8434   ISSN 1617-9692   (electronic)
Lecture Notes in Mathematics
ISBN 978-3-319-46737-5   ISBN 978-3-319-46738-2   (eBook)
DOI 10.1007/978-3-319-46738-2

Library of Congress Control Number: 2016963106

Mathematics Subject Classification (2010): 11-XX; 43-XX; 46-XX; 60-XX; 70-XX

Printed on acid-free paper

This Springer imprint is published by Springer Nature
The registered company is Springer International Publishing AG
The registered company address is: Gewerbestrasse 11, 6330 Cham, Switzerland

Dedicated to my wife Mónica, my daughter Daniela, and my son Felipe

Preface

In recent years, p-adic analysis (or more generally non-Archimedean analysis) has received a lot of attention due to its connections with mathematical physics; see, e.g., [8–14, 20, 21, 25, 26, 36, 64, 65, 67–69, 75, 76, 80, 82, 85–87, 90, 94, 106–109, 111] and references therein. All these developments have been motivated by two physical ideas. The first is the conjecture (due to I. Volovich) in particle physics that at Planck distances the space-time has a non-Archimedean structure; see, e.g., [107, 112, 113]. The second idea comes from statistical physics, in particular, in connection with models describing relaxation in glasses, macromolecules, and proteins. It has been proposed that the non-exponential nature of those relaxations is a consequence of a hierarchical structure of the state space which can in turn be put in connection with p-adic structures; see, e.g., [6, 10–12, 46, 94]. Additionally, we should mention that in the middle of the 1980s the idea of using ultrametric spaces to describe the states of complex biological systems, which naturally possess a hierarchical structure, emerged in the works of Frauenfelder, Parisi, and Stain, among others; see, e.g., [6, 36, 46]. In protein physics, it is regarded as one of the most profound ideas put forward to explain the nature of distinctive life attributes.

On the other hand, stochastic processes on p-adic spaces, or more generally on ultrametric spaces, have been studied extensively in the last 30 years due to diverse physical and mathematical motivations; see, e.g., [1–4, 16, 17, 19, 28, 42, 43, 61–63, 71, 98, 116, 118, 122] and references therein.

In [10–12], Avetisov et al. introduced a new class of models for complex systems based on p-adic analysis. These models can be applied, for instance, to the study of the relaxation of certain biological complex systems. From a mathematical point of view, in these models, the time-evolution of a complex system is described by a p-adic master equation (a parabolic-type pseudodifferential equation) which controls the time-evolution of a transition function of a Markov process on an ultrametric space, and this stochastic process is used to describe the dynamics of the system in the space of configurational states which is approximated by an ultrametric space ($\mathbb{Q}_p$). The first goal of this work is to study a very large class of heat-type pseudodifferential equations over p-adic and adelic spaces, which contains as special cases many of the equations that occur in the models of Avetisov et al. It is worth to mention here that the p-adic heat equation also appeared in certain works connected with the Riemann hypothesis [83].

The simplest type of master equation is the one-dimensional p-adic heat equation. This equation was introduced in the book of Vladimirov, Volovich, and Zelenov [111, Section XVI]. In [80, Chaps. 4 and 5] Kochubei presented a general theory for one-dimensional parabolic-type pseudodifferential equations with variable coefficients, whose fundamental solutions are transition density functions for Markov processes in the p-adic line; see also [97, 98, 108]. Varadarajan studied the heat equation on a division algebra over a non-Archimedean local field [108]. In [122], the author introduced p-adic analogs for the n-dimensional elliptic operators and studied the corresponding heat equations and the associated Markov processes; see also [23, 108]. In [26], building up on [25] and [80, 81], Chacón-Cortés and the author introduced a new type of nonlocal operators and a class of parabolic-type pseudodifferential equations with variable coefficients, which contains the one-dimensional p-adic heat equation of [111], the equations studied by Kochubei in [80], and the equations studied by Rodríguez-Vega in [97].

The field of p-adic numbers $\mathbb{Q}_p$ is defined as the completion of the field of rational numbers $\mathbb{Q}$ with respect to the p-adic norm $|\cdot|_p$; see Chap. 2. The p-adic norm satisfies $||x + y||_p \leq \max\{||x||_p, ||y||_p\}$, and the metric space $(\mathbb{Q}_p^n, ||\cdot||_p)$ is a complete ultrametric space. This space has a natural hierarchical structure, which is very useful in physical models that involve hierarchies. As a topological space, $\mathbb{Q}_p$ is homeomorphic to a Cantor-like subset of the real line.

The p-adic heat equation is defined as

$$\frac{\partial u\,(x, t)}{\partial t} + (\boldsymbol{D}^\alpha u)\,(x, t) = 0, x \in \mathbb{Q}_p, t > 0 \qquad (1)$$

where

$$(\boldsymbol{D}^\alpha \varphi)\,(x) = \mathcal{F}_{\xi \to x}^{-1}\left(|\xi|_p^\alpha \, \mathcal{F}_{x \to \xi}\varphi\right), \alpha > 0$$

is the Vladimirov operator and $\mathcal{F}$ denotes the p-adic Fourier transform. This equation is the p-adic counterpart of the classical heat equation, which describes a particle performing a random motion (the Brownian motion); a 'similar' statement is valid for the p-adic heat equation. More precisely, the fundamental solution of (1) is the transition density of a bounded right-continuous Markov process without second kind discontinuities.

A well-known and accepted scientific paradigm in physics of complex systems (such as glasses and proteins) asserts that the dynamics of a large class of complex systems is described as a random walk on a complex energy landscape; see, e.g., [6, 46, 47, 114], [82] and references therein. A landscape is a continuous real-valued function that represents the energy of a system. The term complex landscape means that energy function has many local minima. In the case of complex landscapes, in which there are many local minima, a "simplification method" called interbasin kinetics is applied. The idea is to study the kinetics generated by transitions between groups of states (basins). In this setting, the minimal basins correspond to local

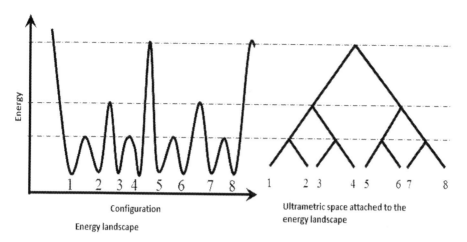

Configuration

Ultrametric space attached to the energy landscape

Energy landscape

Fig. 1

minima of energy, and the large basins (superbasins or union of basins) have a hierarchical structure. A key idea is that a complex landscape is approximated by a disconnectivity graph (an ultrametric space) and by a function on this graph that gives the distribution function of activation energies. For the construction of the disconnectivity graph, we can imagine that the energy landscape is a water tank and that we are pumping water into it. The energy landscape of the system is flooded with water, and water forms pools around the local minima. By increasing the level of water, pools merge, until only one big pool remains. This procedure allows us to construct a directed tree of basins (pools). The activation barrier function on this tree is constructed as follows: minimal pools are assigned their depth (local minima energy), and larger pools are assigned energy levels at which these pools emerge (the activation barrier between basins). This procedure is illustrated in Fig. 1. For "real" energy landscapes and the corresponding graphs, the reader may see [114]. The next step is to construct a model of hierarchical dynamics based on the disconnectivity graph. By using the postulates of the interbasin kinetics, one gets that the transitions between basins are described by the following equations:

$$\frac{\partial f(i,t)}{\partial t} = \sum_j T(j,i) f(j,t) v(j) - \sum_{j \neq i} T(i,j) f(i,t) v(j),$$

where the indices i, j number the states of the system (which correspond to local minima of energy), $T(i,j) \geq 0$ is the probability per unit time of a transition from i to j, and the $v(j) > 0$ are the basin volumes. Under suitable physical and mathematical hypotheses, the above master equation has the following "continuous limit":

$$\frac{\partial f(x,t)}{\partial t} = \int_{\mathbb{Q}_p} [w(x|y) f(y,t) - w(y|x) f(x,t)] \, dy,$$

where $x \in \mathbb{Q}_p, t \geq 0$. The function $f(x, t) : \mathbb{Q}_p \times \mathbb{R}_+ \to \mathbb{R}_+$ is a probability density distribution, so that $\int_B f(x, t) \, dx$ is the probability of finding the system in a domain $B \subset \mathbb{Q}_p$ at the instant t. The function $w(x|y) : \mathbb{Q}_p \times \mathbb{Q}_p \to \mathbb{R}_+$ is the probability of the transition from state y to state x per unit of time. This family of parabolic-type equations contains the p-adic heat equation as a particular case.

The first part of this work is dedicated to present a snapshot of the theory, still under construction, of pseudodifferential equations of parabolic type and their Markov processes on p-adic and adelic spaces. In Chap. 1, we review, without proofs, some basic aspects of p-adic analysis and p-adic manifolds that we use in this work. An interesting novelty of this work is that some of the equations studied here require using geometric methods, for instance, integration on p-adic manifolds. Chapters 2 and 3 are dedicated to the study of parabolic-type equations and their Markov processes. The results presented in these chapters continue and extend, in a considerable form, the corresponding results presented in the books [80] and [111]. Chapter 4 deals with the heat equation on the ring of adeles. We present only the essential techniques and results; many important related topics were left aside. For instance, we do not include pseudodifferential operators and wavelets on general locally compact ultrametric spaces, not necessarily having a group structure; see, e.g., [77, 82] and [66]; wavelet analysis on adeles and pseudodifferential operators; see, e.g., [74] and [73]; and p-adic Brownian motion and stochastic pseudodifferential equations; see, e.g., [21, 70, 71], among other several important topics.

The second part, Chaps. 5 and 6, is dedicated to pseudodifferential equations whose symbols involve polynomials. In the 1950s Gel'fand and Shilov showed that fundamental solutions for certain types of partial differential operators with constant coefficients can be obtained by using local zeta functions [49]. The existence of fundamental solutions for general differential operators with constant coefficients was established by Atiyah [15] and Bernstein [18] using local zeta functions. A similar program can be carried out in the p-adic setting; see, e.g., [120]; see also [123]. The goal of Chap. 5 is to prove the existence of fundamental solutions for pseudodifferential operators via local zeta functions. Chapter 6 deals with a new class of non-Archimedean pseudodifferential equations of Klein-Gordon type. These equations have many similar properties to the classical Klein-Gordon equations; see, e.g., [31, 32, 103]. Finally, in Chap. 7, we present some open problems connecting non-Archimedean pseudodifferential equations with number theory, probability, and physics.

These notes were intended as a one-semester doctoral course at CINVESTAV for students interested in doing research in p-adic analysis connected with physics of complex systems, probability, and number theory. There are many open problems in this area. The central goal of these notes was to prepare fastly my students to do research.

I wish to thank all my coauthors (Oscar F. Casas-Sánchez, Leonardo Chacón-Cortes, Jeanneth Galeano-Peñaloza, John J. Rodríguez-Vega, and Sergii M. Torba) and my current doctoral students and postdocs (Victor Aguilar, Samuel Arias, Edilberto Arroyo, Maria Luisa Mendoza, and Anselmo Torresblanca) for helping me on this project. I also wish to thank my family (my wife Monica, my daughter Daniela, and my son Felipe) for their constant support.

Finally, I wish to thank the CONACYT for supporting my research activities during the last 10 years through several grants (my latest CONACYT grant no. 250845) and through the Mexican Research Chairs Program (SNI level III since 2016).

Mexico   W.A. Zúñiga-Galindo

Contents

Chapter 1
p-Adic Analysis: Essential Ideas and Results

In this chapter, we present, without proofs, the essential aspects and basic results on *p*-adic functional analysis needed in the book. For a detailed exposition on *p*-adic analysis the reader may consult [5, 105, 111].

1.1 The Field of *p*-Adic Numbers

Along this book *p* will denote a prime number. The field of *p*-adic numbers $\mathbb{Q}_p$ is defined as the completion of the field of rational numbers $\mathbb{Q}$ with respect to the *p*-adic norm $|\cdot|_p$, which is defined as

$$|x|_p = \begin{cases} 0 & \text{if } x = 0 \\ p^{-\gamma} & \text{if } x = p^\gamma \dfrac{a}{b}, \end{cases}$$

where *a* and *b* are integers coprime with *p*. The integer $\gamma := ord(x)$, with $ord(0) := +\infty$, is called the *p*-adic order of x. We extend the *p*-adic norm to $\mathbb{Q}_p^n$ by taking

$$||x||_p := \max_{1 \leq i \leq n} |x_i|_p, \qquad \text{for } x = (x_1, \ldots, x_n) \in \mathbb{Q}_p^n.$$

We define $ord(x) = \min_{1 \leq i \leq n}\{ord(x_i)\}$, then $||x||_p = p^{-ord(x)}$. The metric space $\left(\mathbb{Q}_p^n, ||\cdot||_p\right)$ is a complete ultrametric space. As a topological space $\mathbb{Q}_p$ is homeomorphic to a Cantor-like subset of the real line, see e.g. [5, 111].

© Springer International Publishing AG 2016
W.A. Zúñiga-Galindo, *Pseudodifferential Equations Over Non-Archimedean Spaces*, Lecture Notes in Mathematics 2174, DOI 10.1007/978-3-319-46738-2_1

Any *p*-adic number $x \neq 0$ has a unique expansion of the form

$$x = p^{ord(x)} \sum_{j=0}^{\infty} x_j p^j,$$

where $x_j \in \{0, 1, 2, \ldots, p-1\}$ and $x_0 \neq 0$. By using this expansion, we define *the fractional part of $x \in \mathbb{Q}_p$*, denoted $\{x\}_p$, as the rational number

$$\{x\}_p = \begin{cases} 0 & \text{if } x = 0 \text{ or } ord(x) \geq 0 \\ p^{ord(x)} \sum_{j=0}^{-ord(x)-1} x_j p^j & \text{if } ord(x) < 0. \end{cases}$$

In addition, any *p*-adic number $x \neq 0$ can be represented uniquely as $x = p^{ord(x)} ac(x)$ with $|ac(x)|_p = 1$; $ac(x)$ is called the *angular component* of x.

1.1.1 Additive Characters

Set $\chi_p(y) := \exp(2\pi i \{y\}_p)$ for $y \in \mathbb{Q}_p$. The map $\chi_p(\cdot)$ is an additive character on $\mathbb{Q}_p$, i.e. a continuous map from $(\mathbb{Q}_p, +)$ into S (the unit circle considered as multiplicative group) satisfying $\chi_p(x_0 + x_1) = \chi_p(x_0)\chi_p(x_1)$, $x_0, x_1 \in \mathbb{Q}_p$. We notice that χ_p satisfies the following relations:

$$\chi_p(0) = 1, \qquad \chi_p(-x) = \overline{\chi_p(x)} = \chi_p^{-1}(x)$$

$$\chi_p(nx) = \chi_p^n(x), \ n \in \mathbb{Z}.$$

The additive characters of $\mathbb{Q}_p$ form an Abelian group which is isomorphic to $(\mathbb{Q}_p, +)$; the isomorphism is given by $\xi \to \chi_p(\xi x)$, see e.g. [5, Section 2.3].

1.1.2 Multiplicative Characters

A multiplicative character of $(\mathbb{Q}_p^\times, \cdot)$, or simply of $\mathbb{Q}_p$, is a continuous mapping π from $(\mathbb{Q}_p^\times, \cdot)$ into $(\mathbb{C}^\times, \cdot)$ satisfying $\pi(x_0 x_1) = \pi(x_0)\pi(x_1)$, $x_0, x_1 \in \mathbb{Q}_p^\times$. Every multiplicative character of $\mathbb{Q}_p$ can be represented as

$$\pi(x) := \pi_s(x) = |x|_p^{s-1} \pi_1(ac(x)),$$

where $s \in \mathbb{C}$ and π_1 is the restriction of π to $\mathbb{Z}_p^\times = \{x \in \mathbb{Q}_p^\times ; |x|_p = 1\}$ into S, see e.g. [5, Section 2.3].

1.2 Topology of $\mathbb{Q}_p^n$

For $r \in \mathbb{Z}$, denote by $B_r^n(a) = \{x \in \mathbb{Q}_p^n; ||x - a||_p \leq p^r\}$ *the ball of radius p^r with center at* $a = (a_1, \ldots, a_n) \in \mathbb{Q}_p^n$, and take $B_r^n(0) := B_r^n$. Note that $B_r^n(a) = B_r(a_1) \times \cdots \times B_r(a_n)$, where $B_r(a_i) := \{x \in \mathbb{Q}_p; |x_i - a_i|_p \leq p^r\}$ is the one-dimensional ball of radius p^r with center at $a_i \in \mathbb{Q}_p$. The ball B_0^n equals the product of n copies of $B_0 = \mathbb{Z}_p$, *the ring of p-adic integers*. We also denote by $S_r^n(a) = \{x \in \mathbb{Q}_p^n; ||x - a||_p = p^r\}$ *the sphere of radius p^r with center at* $a = (a_1, \ldots, a_n) \in \mathbb{Q}_p^n$, and take $S_r^n(0) := S_r^n$. We notice that $S_0^1 = \mathbb{Z}_p^\times$ (the group of units of $\mathbb{Z}_p$), but $\left(\mathbb{Z}_p^\times\right)^n \subsetneq S_0^n$. The balls and spheres are both open and closed subsets in $\mathbb{Q}_p^n$. In addition, two balls in $\mathbb{Q}_p^n$ are either disjoint or one is contained in the other.

As a topological space $\left(\mathbb{Q}_p^n, || \cdot ||_p\right)$ is totally disconnected, i.e. the only connected subsets of $\mathbb{Q}_p^n$ are the empty set and the points. A subset of $\mathbb{Q}_p^n$ is compact if and only if it is closed and bounded in $\mathbb{Q}_p^n$, see e.g. [111, Section 1.3], or [5, Section 1.8]. The balls and spheres are compact subsets. Thus $\left(\mathbb{Q}_p^n, || \cdot ||_p\right)$ is a locally compact topological space.

We will use $\Omega\left(p^{-r}||x - a||_p\right)$ to denote the characteristic function of the ball $B_r^n(a)$. We will use the notation 1_A for the characteristic function of a set A.

1.3 The Bruhat-Schwartz Space and the Fourier Transform

A complex-valued function φ defined on $\mathbb{Q}_p^n$ is *called locally constant* if for any $x \in \mathbb{Q}_p^n$ there exist an integer $l(x) \in \mathbb{Z}$ such that

$$\varphi(x + x') = \varphi(x) \text{ for } x' \in B_{l(x)}^n. \tag{1.1}$$

The $\mathbb{C}$-vector space of locally constant functions will be denoted as $\mathcal{E}(\mathbb{Q}_p^n)$. The locally constant functions φ for which the integer $l(x)$ in (1.1) depends only on φ form a $\mathbb{C}$-vector subspace of $\mathcal{E}(\mathbb{Q}_p^n)$ denoted as $\tilde{\mathcal{E}}(\mathbb{Q}_p^n)$.

A function $\varphi : \mathbb{Q}_p^n \to \mathbb{C}$ is called a *Bruhat-Schwartz function (or a test function)* if it is locally constant with compact support. Any test function can be represented as a linear combination, with complex coefficients, of characteristic functions of balls. The $\mathbb{C}$-vector space of Bruhat-Schwartz functions is denoted by $\mathcal{D}(\mathbb{Q}_p^n) := \mathcal{D}$. For $\varphi \in \mathcal{D}(\mathbb{Q}_p^n)$, the largest number $l = l(\varphi)$ satisfying (1.1) is called *the exponent of local constancy (or the parameter of constancy) of* φ. We warn the reader about the fact that slightly different notions of exponent of local constancy will be used in Chaps. 3 and 4.

Convergence in $\mathcal{D}(\mathbb{Q}_p^n)$ is defined in the following way: $\varphi_k \to 0$, $k \to \infty$, in $\mathcal{D}(\mathbb{Q}_p^n)$ if and only if: (i) $\operatorname{supp}\varphi_k \subset B_N^n$ with N independent of k, and $l(\varphi_k) \geq l$ with l independent of k; (ii) $\varphi_k \xrightarrow{\quad} \text{uniformly } 0 \text{ in } \mathbb{Q}_p^n$.

The space $\mathcal{D}(\mathbb{Q}_p^n)$ is a complete locally convex topological vector space. If U is an open subset of $\mathbb{Q}_p^n$, $\mathcal{D}(U)$ denotes the space of test functions with supports contained in U, then $\mathcal{D}(U)$ is dense in

$$L^\rho(U) = \left\{ \varphi : U \to \mathbb{C}; \|\varphi\|_\rho = \left\{ \int |\varphi(x)|^\rho \, d^n x \right\}^{\frac{1}{\rho}} < \infty \right\},$$

where $d^n x$ is the Haar measure on $\mathbb{Q}_p^n$ normalized by the condition $vol(B_0^n) = 1$, for $1 \le \rho < \infty$, see e.g. [5, Section 4.3]. We will also use the simplified notation L^ρ, $1 \le \rho < \infty$, if there is no danger of confusion.

1.3.1   The Fourier Transform of Test Functions

Given $\xi = (\xi_1, \dots, \xi_n)$ and $y = (x_1, \dots, x_n) \in \mathbb{Q}_p^n$, we set $\xi \cdot x := \sum_{j=1}^n \xi_j x_j$. The Fourier transform of $\varphi \in \mathcal{D}(\mathbb{Q}_p^n)$ is defined as

$$(\mathcal{F}\varphi)(\xi) = \int_{\mathbb{Q}_p^n} \chi_p(\xi \cdot x)\varphi(x) d^n x \quad \text{for } \xi \in \mathbb{Q}_p^n,$$

where $d^n x$ is the normalized Haar measure on $\mathbb{Q}_p^n$. The Fourier transform is a linear isomorphism from $\mathcal{D}(\mathbb{Q}_p^n)$ onto itself satisfying $(\mathcal{F}(\mathcal{F}\varphi))(\xi) = \varphi(-\xi)$, see e.g. [5, Section 4.8]. We will also use the notation $\mathcal{F}_{x\to\xi}\varphi$ and $\hat{\varphi}$ for the Fourier transform of φ.

In the definition of the Fourier transform, the bilinear form $\xi \cdot x$ can be replaced for any symmetric non-degenerate bilinear form $[\xi, x]$. We will use such Fourier transforms in Chap. 6.

1.4   Distributions

Let $\mathcal{D}'(\mathbb{Q}_p^n)$ denote the $\mathbb{C}$-vector space of all continuous functionals (distributions) on $\mathcal{D}(\mathbb{Q}_p^n)$. The natural pairing $\mathcal{D}'(\mathbb{Q}_p^n) \times \mathcal{D}(\mathbb{Q}_p^n) \to \mathbb{C}$ is denoted as (T, φ) for $T \in \mathcal{D}'(\mathbb{Q}_p^n)$ and $\varphi \in \mathcal{D}(\mathbb{Q}_p^n)$. Convergence in $\mathcal{D}'(\mathbb{Q}_p^n)$ is defined as the weak convergence: $T_k \to 0$, $k \to \infty$, in $\mathcal{D}'(\mathbb{Q}_p^n)$ if $(T_k, \varphi) \to 0$, $k \to \infty$, for all $\varphi \in \mathcal{D}(\mathbb{Q}_p^n)$. The space $\mathcal{D}'(\mathbb{Q}_p^n)$ agrees with the algebraic dual of $\mathcal{D}(\mathbb{Q}_p^n)$, i.e. all functionals on $\mathcal{D}(\mathbb{Q}_p^n)$ are continuous. In addition, $\mathcal{D}'(\mathbb{Q}_p^n)$ is complete, i.e. if $T_k - T_j \to 0$, $k, j \to \infty$, then there exits a functional $T \in \mathcal{D}'(\mathbb{Q}_p^n)$ such that $T_k - T \to 0, k \to \infty$ in $\mathcal{D}'(\mathbb{Q}_p^n)$, see e.g. [5, Section 4.4].

Let U be an open subset of $\mathbb{Q}_p^n$. A distribution $T \in \mathcal{D}'(U)$ vanishes on $V \subset U$ if $(T, \varphi) = 0$ for all $\varphi \in \mathcal{D}(V)$. Let $U_T \subset U$ be the maximal open subset on which the distribution T vanishes. The support of T is the complement of U_T in U. We denote it by $\operatorname{supp} T$.

Given a fixed test function θ and a distribution $T \in \mathcal{D}'\left(\mathbb{Q}_p^n\right)$, we define the distribution θT by the formula $(\theta T, \varphi) = (T, \theta\varphi)$ for $\varphi \in \mathcal{D}\left(\mathbb{Q}_p^n\right)$. We say that a distribution $T \in \mathcal{D}'\left(\mathbb{Q}_p^n\right)$ has *compact support* if there is $k \in \mathbb{Z}$ such that $\Delta_k T = T$ in $\mathcal{D}'\left(\mathbb{Q}_p^n\right)$, where $\Delta_k(x) := \Omega\left(p^{-k}\|x\|_p\right)$.

Every $f \in \mathcal{E}\left(\mathbb{Q}_p^n\right)$, or more generally in L^1_{loc}, defines a distribution $f \in \mathcal{D}'\left(\mathbb{Q}_p^n\right)$ by the formula

$$(f, \varphi) = \int_{\mathbb{Q}_p^n} f(x)\,\varphi(x)\,d^n x.$$

Such distributions are called *regular distributions*.

1.4.1 The Fourier Transform of a Distribution

The Fourier transform $\mathcal{F}[T]$ of a distribution $T \in \mathcal{D}'\left(\mathbb{Q}_p^n\right)$ is defined by

$$(\mathcal{F}[T], \varphi) = (T, \mathcal{F}[\varphi]) \text{ for all } \varphi \in \mathcal{D}\left(\mathbb{Q}_p^n\right).$$

The Fourier transform $T \to \mathcal{F}[T]$ is a linear (and continuous) isomorphism from $\mathcal{D}'\left(\mathbb{Q}_p^n\right)$ onto $\mathcal{D}'\left(\mathbb{Q}_p^n\right)$. Furthermore, $T = \mathcal{F}[\mathcal{F}[T](-\xi)]$.

Let $T \in \mathcal{D}'\left(\mathbb{Q}_p^n\right)$ be a distribution. Then $\operatorname{supp} T \subset B_N^n$ if and only if $\mathcal{F}[T] \in \tilde{\mathcal{E}}(\mathbb{Q}_p^n)$, where the exponent of local constancy of $\mathcal{F}[T]$ is $\geq -N$. In addition

$$\mathcal{F}[T](\xi) = \left(T(y), \Delta_N(y)\chi_p(\xi \cdot y)\right),$$

see e.g. [5, Section 4.9].

1.4.2 The Direct Product of Distributions

Given $F \in \mathcal{D}'\left(\mathbb{Q}_p^n\right)$ and $G \in \mathcal{D}'\left(\mathbb{Q}_p^m\right)$, their *direct product* $F \times G$ is defined by the formula

$$(F(x) \times G(y), \varphi(x, y)) = (F(x), (G(y), \varphi(x, y))) \text{ for } \varphi(x, y) \in \mathcal{D}\left(\mathbb{Q}_p^{n+m}\right).$$

The direct product is commutative: $F \times G = G \times F$. In addition the direct product is continuous with respect to the joint factors.

1.4.3   The Convolution of Distributions

Given $F, G \in \mathcal{D}'\left(\mathbb{Q}_p^n\right)$, their convolution $F * G$ is defined by

$$(F * G, \varphi) = \lim_{k \to \infty} (F(y) \times G(x), \Delta_k(x) \varphi(x+y))$$

if the limit exists for all $\varphi \in \mathcal{D}\left(\mathbb{Q}_p^n\right)$. We recall that if $F * G$ exists, then $G * F$ exists and $F * G = G * F$, see e.g. [111, Section 7.1]. If $F, G \in \mathcal{D}'\left(\mathbb{Q}_p^n\right)$ and $\mathrm{supp}\, G \subset B_N^n$, then the convolution $F * G$ exists, and it is given by the formula

$$(F * G, \varphi) = (F(y) \times G(x), \Delta_N(x) \varphi(x+y)) \ \text{ for } \varphi \in \mathcal{D}\left(\mathbb{Q}_p^n\right).$$

In the case in which $G = \psi \in \mathcal{D}\left(\mathbb{Q}_p^n\right)$, $F * \psi$ is a locally constant function given by

$$(F * \psi)(y) = (F(x), \psi(y-x)),$$

see e.g. [111, Section 7.1].

1.4.4   The Multiplication of Distributions

Set $\delta_k(x) := p^{nk} \Omega\left(p^k \|x\|_p\right)$ for $k \in \mathbb{N}$. Given $F, G \in \mathcal{D}'\left(\mathbb{Q}_p^n\right)$, their product $F \cdot G$ is defined by

$$(F \cdot G, \varphi) = \lim_{k \to \infty} (G, (F * \delta_k)\varphi)$$

if the limit exists for all $\varphi \in \mathcal{D}\left(\mathbb{Q}_p^n\right)$. If the product $F \cdot G$ exists then the product $G \cdot F$ exists and they are equal.

We recall that the existence of the product $F \cdot G$ is equivalent to the existence of $\mathcal{F}[F] * \mathcal{F}[G]$. In addition, $\mathcal{F}[F \cdot G] = \mathcal{F}[F] * \mathcal{F}[G]$ and $\mathcal{F}[F * G] = \mathcal{F}[F] \cdot \mathcal{F}[G]$, see e.g. [111, Section 7.5].

1.5   Essential Aspects of the *p*-Adic Manifolds

In this section we review, without proofs, some results and techniques about *p*-adic manifolds (in the sense of Serre), that we will use later on in the study of certain pseudodifferential equations. The material is based on Igusa's book [57], the reader may consult this reference for further details.

1.5.1 Implicit Function Theorem

Let us denote by $\mathbb{Q}_p[[x_1,\ldots,x_n]]$, the *ring of formal power series* with coefficients in $\mathbb{Q}_p$. An element of this ring has the form

$$\sum c_i x^i = \sum_{(i_1,\ldots,i_n)\in\mathbb{N}^n} c_{i_1,\ldots,i_n} x_1^{i_1}\ldots x_n^{i_n}.$$

A formal series $\sum c_i x^i$ is said to be *convergent* if there exists $r \in \mathbb{Z}$ such that $\sum c_i a^i$ converges for any $a = (a_1,\ldots,a_n) \in \mathbb{Q}_p^n$ satisfying $\|a\|_p < p^r$. The convergent series form a subring of $\mathbb{Q}_p[[x_1,\ldots,x_n]]$, which will be denoted as $\mathbb{Q}_p\langle\langle x_1,\ldots,x_n\rangle\rangle$.

If for $\sum c_i x^i$ there exists $\sum c_i^{(0)} x^i \in \mathbb{R}\langle\langle x_1,\ldots,x_n\rangle\rangle$ such that $|c_i|_p \leq c_i^{(0)}$ for all $i \in \mathbb{N}^n$, we say that $\sum c_i^{(0)} x^i$ is a *dominant series* for $\sum c_i x^i$. A formal power series is convergent if and only if it has a dominant series, see [57, Lemma 2.1.1].

Theorem 1

(i) *Take* $F(x,y) = (F_1(x,y),\ldots,F_m(x,y))$, *with* $F_i(x,y) \in \mathbb{Q}_p[[x_1,\ldots,x_n,y_1,\ldots,y_m]]$ *such that* $F_i(0,0) = 0$, *and*

$$\det\left[\frac{\partial F_i}{\partial y_j}(0,0)\right]_{\substack{1\leq i\leq m\\1\leq j\leq m}} \neq 0.$$

Then there exists a unique $f(x) = (f_1(x),\ldots,f_m(x))$, *with* $f_i(0) = 0$, $f_i(x) \in \mathbb{Q}_p[[x_1,\ldots,x_n]]$, *satisfying* $F(x,f(x)) = 0$, *i.e.* $F_i(x,f(x)) = 0$ *for all* i.

(ii) *If each* $F_i(x,y)$ *is a convergent power series, then every* $f_i(x)$ *is a convergent power series. Furthermore if* a *is near* 0 *in* $\mathbb{Q}_p^n$, *then* $f(a)$ *is near* 0 *in* $\mathbb{Q}_p^n$ *and* $F(a,f(a)) = 0$, *and if* (a,b) *is near* $(0,0)$ *in* $\mathbb{Q}_p^n \times \mathbb{Q}_p^m$ *and* $F(a,b) = 0$, *then* $b = f(a)$.

For the proof the reader may consult [57, Theorem 2.1.1].

Corollary 2

(i) *If* $g_i(x) \in \mathbb{Q}_p[[x_1,\ldots,x_n]]$, $g_i(0) = 0$ *for* $1 \leq i \leq n$, *and*

$$\det\left[\frac{\partial g_i}{\partial x_j}(0)\right] \neq 0,$$

then there exists a unique $f(x) = (f_1(x),\ldots,f_m(x))$ *with* $f_i(0) = 0$, $f_i(x) \in \mathbb{Q}_p[[x_1,\ldots,x_n]]$, *for all* i, *such that* $g(f(x)) = x$.

(ii) *If* $g_i(x) \in \mathbb{Q}_p\langle\langle x_1,\ldots,x_n\rangle\rangle$, *then* $f_i(x) \in \mathbb{Q}_p\langle\langle x_1,\ldots,x_n\rangle\rangle$ *for all* i. *Furthermore if* b *is near* 0 *in* $\mathbb{Q}_p^n$ *and* $a = g(b)$, *then* a *is also near* 0 *in* $\mathbb{Q}_p^n$ *and* $b = f(a)$. *Therefore* $y = f(x)$ *gives rise to a bicontinuous map from a small neighborhood of* 0 *in* $\mathbb{Q}_p^n$ *to another* $\mathbb{Q}_p^n$.

Remark 3 Take $U_1 \subset \mathbb{Q}_p^n$, $U_2 \subset \mathbb{Q}_p^m$, open subsets containing the origin. Assume that each $F_i(x, y) : U_1 \times U_2 \to \mathbb{Q}_p$ is a convergent power series. A set of the form

$$V := \{(x, y) \in U_1 \times U_2; F_i(x, y), i = 1, \ldots, m\},$$

is called an *analytic set*. In the case in which all the $F_i(x, y)$ are polynomials and $U_1 = \mathbb{Q}_p^n$, $U_2 = \mathbb{Q}_p^m$, V is called an *algebraic set*. If all the $F_i(x, y) \in \mathbb{Q}_p\langle\langle x, y \rangle\rangle$ satisfy the hypotheses of the implicit function theorem, V has a parametrization, possible after shrinking U_1, U_2, i.e. there exist open subsets containing the origin $\tilde{U}_1 \subset U_1$, $\tilde{U}_2 \subset U_2$, such that

$$V = \{(x, y) \in \tilde{U}_1 \times \tilde{U}_2; F_i(x, y) = 0, i = 1, \ldots, m\}$$

$$= \{(x, y) \in \tilde{U}_1 \times \tilde{U}_2; y = f(x)\}.$$

If we now use as coordinates

$$x_1, \ldots, x_n, z_1 = y_1 - f_1(x), \ldots, z_m = y_m - f_m(x),$$

we have

$$V = \{(x, z) \in \tilde{U}_1 \times \tilde{U}_2; z_1 = \ldots = z_m = 0\}.$$

We say that V is *a closed analytic submanifold of $\tilde{U}_1 \times \tilde{U}_2 \subset \mathbb{Q}_p^n \times \mathbb{Q}_p^m$ of codimension m*. The word 'closed' means that V is closed in the *p*-adic topology.

1.5.2 *p*-Adic Analytic Manifolds

Let $U \subset \mathbb{Q}_p^n$ be a non-empty open set, and let $f : U \to \mathbb{Q}_p$ be a function. If at every point $a = (a_1, \ldots, a_n)$ of U there exists an element $f_a(x) \in \mathbb{Q}_p\langle\langle x - a \rangle\rangle = \mathbb{Q}_p\langle\langle x_1 - a_1, \ldots, x_n - a_n \rangle\rangle$ such that $f(x) = f_a(x)$ for any point x near to a, we say that f is an *analytic function on U*. It is not hard to show that all the partial derivatives of f are analytic on U.

Let U be as above and let $h = (h_1, \ldots, h_m) : U \to \mathbb{Q}_p^m$ be a mapping. If each h_i is an analytic function on U, we say that h is *an analytic mapping on U*.

Let X denote a Hausdorff space and n a fixed nonnegative integer. A pair (U, ϕ_U), where U is a nonempty open subset of X and $\phi_U : U \to \phi_U(U)$ is a bicontinuous map (i.e. a homeomorphism) from U to the open set $\phi_U(U)$ of $\mathbb{Q}_p^n$, is called *a chart*. Furthermore $\phi_U(x) = (x_1, \ldots, x_n)$ (for a variable point x of U) are called *the local coordinates* of x. A set of charts $\{(U, \phi_U)\}$ is called an *atlas* if the union all U is X and for every U, U' such that $U \cap U' \neq \emptyset$ the map

$$\phi_{U'} \circ \phi_U^{-1} : \phi_U(U \cap U') \to \phi_{U'}(U \cap U')$$

is analytic. Two atlases are considered equivalent if their union is also an atlas. This is an equivalence relation and any equivalence class is called an *n-dimensional p-adic analytic structure on X*. If $\{(U, \phi_U)\}$ is an atlas, we say that X is an *n-dimensional p-adic analytic manifold,* and we write $n = \dim(X)$.

Suppose that X, Y are *p*-adic analytic manifolds respectively defined by $\{(U, \phi_U)\}$, $\{(V, \psi_V)\}$ and $f : X \rightarrow Y$ is a map. If for every U, V such that $U \cap f^{-1}(V) \neq \emptyset$, the map

$$\psi_V \circ f \circ \phi_U^{-1} : \phi_U\left(U \cap f^{-1}(V)\right) \rightarrow \mathbb{Q}_p^{\dim(Y)}$$

is analytic, then we say that f is an *analytic map.* This notion does not depend on the choice of atlases.

Suppose that X is a *p*-adic analytic manifold defined by $\{(U, \phi_U)\}$ and Y is a nonempty open subset of X. If for every $U' = Y \cap U \neq \emptyset$ we put $\phi_{U'} = \phi_U \mid_{U'}$, then $\{(U', \phi_{U'})\}$ gives an atlas on Y, which makes Y a *p*-adic analytic *open submanifold of X,* with $\dim(X) = \dim(Y)$.

Suppose that Y is a nonempty closed subset of X, a *p*-adic analytic manifold as before, and $0 < m \leq n$ such that an atlas $\{(U, \phi_U)\}$ defining X can be chosen with the following property: if $\phi_U(x) = (x_1, \ldots, x_n)$ and $U' = Y \cap U \neq \emptyset$, then there exist analytic functions $F_1, \ldots, F_m$ on U such that firstly U' becomes the set of all x in U satisfying $F_1(x) = \ldots = F_m(x) = 0$ and secondly

$$\det\left[\frac{\partial F_i}{\partial x_j}\right]_{\substack{1 \leq i \leq m \\ 1 \leq j \leq m}} (a) \neq 0 \text{ at every } a \text{ in } U'.$$

Then by Corollary 2 (ii) the mapping

$$x \rightarrow (F_1(x), \ldots, F_m(x), x_{m+1}, \ldots, x_n)$$

is an analytic mapping from a neighborhood of a in U to its image in $\mathbb{Q}_p^n$. Now, if we denote by V the intersection of such neighborhood of a and Y, and put $\psi_V(x) = (x_{m+1}, \ldots, x_n)$ for every x in V, then $\{(V, \psi_V)\}$ gives an atlas on Y. Therefore Y becomes a *p*-adic analytic manifold with $\dim(Y) = n - m$. We call Y a *closed submanifold of X of codimension m.*

1.5.3   Integration on p-Adic Manifolds

Let μ_n denote the normalized Haar measure of $\mathbb{Q}_p^n$. Take X and $\{(U, \phi_U)\}$ as before. Set α a differential form of degree n on X, then $\alpha \mid U$ has an expression of the form

$$\alpha(x) = f_U(x)\, dx_1 \wedge \ldots \wedge dx_n,$$

in which is f_U an analytic function on U. If A is an open and compact subset of X contained in U, then we define its measure $\mu_\alpha(A)$ as

$$
\mu_\alpha(A) = \int_A |f_U(x)|_p \, \mu_n(\phi_U(x))
$$

$$
= \sum_{e \in \mathbb{Z}} p^{-e} \mu_n\left(\phi_U\left(f_U^{-1}\left(p^e \mathbb{Z}_p^\times\right) \cap A\right)\right). \tag{1.2}
$$

We note that the above series converges because $f_U(A)$ is a compact subset. If $(U', \phi_{U'})$ is another chart and $A \subset U'$, then we will have the same $\mu_\alpha(A)$ relative to that chart. In fact, if $\phi_{U'}(x) = (x_1', \ldots, x_n') = x'$, then

$$
f_{U'}(x') \, dx_1' \wedge \ldots \wedge dx_n' = f_{U'}(\phi_{U'}(x)) \det\left[\frac{\partial x_i'}{\partial x_i}\right] dx_1 \wedge \ldots \wedge dx_n
$$

$$
= f_U(x) \, dx_1 \wedge \ldots \wedge dx_n,
$$

i.e.

$$
f_U(x) = f_{U'}(\phi_{U'}(x)) \det\left[\frac{\partial x_i'}{\partial x_i}\right],
$$

$$
\mu_n(\phi_U(x)) = \left|\det\left[\frac{\partial x_i'}{\partial x_i}\right]\right|_p \mu_n(\phi_{U'}(x'))
$$

which is the rule for changing variables in (1.2):

$$
\int_A |f_U(x)|_p \, \mu_n(\phi_U(x)) = \int_A |f_{U'}(\phi_{U'}(x))|_p \left|\det\left[\frac{\partial x_i'}{\partial x_i}\right]\right|_p \mu_n(\phi_{U'}(x')),
$$

see [57, p. 112 and Proposition 7.4.1]. Note that if $X = U \subset \mathbb{Q}_p^n$ and $\alpha = dx_1 \wedge \ldots \wedge dx_n$ then μ_α is the normalized Haar measure of $\mathbb{Q}_p^n$, which is being denoted as $d^n x$. For further details the reader may consult [57, 104].

1.5.4   Integration Over the Fibers

We now recall some basic facts about the integration over fibers (see e.g. [57, Sect. 7.6]). We denote by $\omega = d\xi_1 \wedge \ldots \wedge d\xi_n$ a differential form of degree n on $\mathbb{Q}_p^n$. Assume that $f : \mathbb{Q}_p^n \to \mathbb{Q}_p$ is a polynomial map such that $C_f \subset f^{-1}(0)$, where

$$C_f := \left\{ x \in \mathbb{Q}_p^n; \nabla f(x) = 0 \right\}$$

is the *critical set* of f. Notice that the hypothesis about the critical set of f implies that

$$V_\lambda := \left\{ x \in \mathbb{Q}_p^n; f(x) = \lambda \right\} \text{ for } \lambda \in \mathbb{Q}_p \smallsetminus \{0\}$$

is a closed submanifold of $\mathbb{Q}_p^n$ of codimension 1. Since df does not vanish on $\mathbb{Q}_p^n \setminus \{0\}$, there exists a $(n-1)$-degree differential form γ_{GL}, called a *Gelfand-Leray form*, such that $\omega = df \wedge \gamma_{\mathrm{GL}}$, which does not vanish on V_λ, $\lambda \in \mathbb{Q}_p^n \setminus \{0\}$. The Gel'fand-Leray form is not unique, but its restriction to V_λ is independent of the choice of γ_{GL}, see [49, Chap. III, Sect. 1–9] and [121]. We denote by $|\gamma_{\mathrm{GL}}|$ the measure induced by the form γ_{GL}. Then

$$\mathcal{D}\left(\mathbb{Q}_p^{n+1}\right) \to \qquad \mathbb{R}$$

$$\phi \qquad \to \int_{V_\lambda} \phi(\xi, \tau) |\gamma_{\mathrm{GL}}|$$

is a measure supported on V_h, in addition,

$$\int_{\mathbb{Q}_p^n} \varphi(\xi) \, d\xi = \int_{\mathbb{Q}_p \setminus \{0\}} \left(\int_{f(\xi) = \lambda} \varphi(\xi) |\gamma_{\mathrm{GL}}| \right) d\lambda, \text{ for } \varphi \in \mathcal{D}\left(\mathbb{Q}_p^n\right), \tag{1.3}$$

(cf. [57, Lemma 8.3.2]).

Chapter 2
Parabolic-Type Equations and Markov Processes

2.1 Introduction

During the last 30 years there have been a strong interest on stochastic processes on ultrametric spaces mainly due its connections with models of complex systems, such as glasses and proteins. These processes are very convenient for describing phenomena whose space of states display a hierarchical structure, see e.g. [9–13, 36, 61, 78, 80, 86, 94, 108, 111, 118, 122], and references therein. Avetisov et al. constructed a wide variety of models of ultrametric diffusion constrained by hierarchical energy landscapes, see [9–13]. From a mathematical point view, in these models the time-evolution of a complex system is described by a p-adic master equation (a parabolic-type pseudodifferential equation) which controls the time-evolution of a transition density function of a Markov process on an ultrametric space. This process describes the dynamics of the system in the space of configurational states which is approximated by an ultrametric space ($\mathbb{Q}_p$). This is the main motivation for developing a general theory of parabolic-type pseudodifferential equations.

This chapter is devoted to the study of several types of n-dimensional parabolic-type equations that are generalizations of the one dimensional p-adic heat equation introduced in [111]. We also study some basic properties of the Markov processes associated with these equations. In Sect. 2.2, we introduce the operators $\boldsymbol{W}$ which are generalizations of the Vladimirov and Taibleson operators. This type of operators was introduced by Chacón-Cortes and Zúñiga-Galindo in [25]. The $\boldsymbol{W}$ operators are pseudodifferential operators having radial symbols. We attach to these symbols certain heat kernels, and show that they are transition density functions of Markov processes over $\mathbb{Q}_p^n$. We also study the Cauchy problem for the parabolic-type equations attached to operators $\boldsymbol{W}$ by using semigroup theory. In Sect. 2.3, we introduce a class of elliptic pseudodifferential operators which are generalizations of the Vladimirov and Taibleson operators. This class of operators was introduced

© Springer International Publishing AG 2016
W.A. Zúñiga-Galindo, *Pseudodifferential Equations Over
Non-Archimedean Spaces*, Lecture Notes in Mathematics 2174,
DOI 10.1007/978-3-319-46738-2_2

by Zúñiga-Galindo in [122]. The symbol of an elliptic operator has the form $|f|_p^\beta$, with $\beta > 0$, where f is a polynomial that vanishes only at the origin. These symbols, in general, are not radial. We attach heat kernels to elliptic symbols and show that these heat kernels are transition density functions of Markov over $\mathbb{Q}_p^n$. The positivity and the decay at infinity of these heat kernels are delicate matters. Finally, we study the Cauchy problem for the heat equations attached to elliptic operators.

2.2 Operators W, Parabolic-Type Equations and Markov Processes

2.2.1 A Class of Non-local Operators

Take $\mathbb{R}_+ := \{x \in \mathbb{R}; x \geq 0\}$, and fix a function

$$w : \mathbb{Q}_p^n \to \mathbb{R}_+$$

satisfying the following properties:

(i) $w(y)$ is a radial (i.e. $w(y) = w(\|y\|_p)$), continuous and increasing function of $\|y\|_p$;
(ii) $w(y) = 0$ if and only if $y = 0$;
(iii) there exist constants $C_0 > 0$, $M \in \mathbb{Z}$, and $\alpha_1 > n$ such that

$$C_0 \|y\|_p^{\alpha_1} \leq w(\|y\|_p), \text{ for } \|y\|_p \geq p^M. \tag{2.1}$$

Note that condition (iii) implies that

$$\int\limits_{\|y\|_p \geq p^M} \frac{d^n y}{w(\|y\|_p)} < \infty. \tag{2.2}$$

In addition, since $w(y)$ is a continuous function, (2.2) holds for any $M \in \mathbb{Z}$.

We define

$$(W\varphi)(x) = \kappa \int\limits_{\mathbb{Q}_p^n} \frac{\varphi(x - y) - \varphi(x)}{w(y)} d^n y, \text{ for } \varphi \in \mathcal{D}, \tag{2.3}$$

where κ is a positive constant.

Lemma 4 *For* $1 \leq \rho \leq \infty$,

$$\mathcal{D}\left(\mathbb{Q}_p^n\right) \to L^\rho\left(\mathbb{Q}_p^n\right)$$

$$\varphi \quad \to \quad W\varphi$$

is a well-defined linear operator. Furthermore,

$$
\mathcal{F}\left[W\varphi\right](\xi) = -\kappa \left(\int\limits_{\mathbb{Q}_p^n} \frac{1 - \chi_p\,(y \cdot \xi)}{w\,(y)} d^n y \right) \mathcal{F}\left[\varphi\right](\xi). \tag{2.4}
$$

Proof Note that

$$
(W\varphi)(x) = \kappa \frac{1_{\mathbb{Q}_p^n \setminus B_{p^M}^n}(x)}{w\,(x)} * \varphi\,(x) - \kappa\varphi\,(x) \left(\int\limits_{\|y\|_p \ge p^M} \frac{d^n y}{w\,(y)} \right), \tag{2.5}
$$

for some constant $M = M(\varphi)$. Now, since $\varphi \in \mathcal{D} \subset L^\rho$, for $1 \le \rho \le \infty$, (2.2), the Young inequality implies that the first term on the right-hand side of (2.5) belongs to L^ρ for $1 \le \rho \le \infty$, and by (2.2) the second term in (2.5) also belongs to L^ρ for $1 \le \rho \le \infty$. Finally, formula (2.4) follows from Fubini's theorem, since

$$
\left| \frac{\varphi\,(x - y) - \varphi\,(x)}{w\,(y)} \right| \in L^1\left(\mathbb{Q}_p^n \times \mathbb{Q}_p^n, d^n x d^n y \right).
$$

∎

We set

$$
A_w\,(\xi) := \int\limits_{\mathbb{Q}_p^n} \frac{1 - \chi_p\,(y \cdot \xi)}{w\,(y)} d^n y.
$$

Lemma 5 *The function $A_w\,(\xi)$ has the following properties: (i) for $\|\xi\|_p = p^{-\gamma} \neq 0$, with $\gamma = ord(\xi)$,*

$$
A_w\left(p^{-\gamma}\right) = (1 - p^{-n}) \sum_{j=\gamma+2}^{\infty} \frac{p^{nj}}{w(p^j)} + \frac{p^{n\gamma + n}}{w(p^{\gamma+1})}; \tag{2.6}
$$

(ii) it is radial, positive, continuous, and $A_w\,(0) = 0$, (iii) $A_w\left(p^{-ord(\xi)}\right)$ is a decreasing function of $ord(\xi)$.

Proof We write $\xi = p^\gamma \xi_0$, with $\gamma = ord(\xi)$ and $\|\xi_0\|_p = 1$. Then

$$
A_w\,(\xi) = \int\limits_{\mathbb{Q}_p^n} \frac{1 - \chi_p\,(p^\gamma y \cdot \xi_0)}{w\,(\|y\|_p)} d^n y = p^{\gamma n} \int\limits_{\mathbb{Q}_p^n} \frac{1 - \chi_p\,(z \cdot \xi_0)}{w\,(p^\gamma\,\|z\|_p)} d^n z. \tag{2.7}
$$

We now note that

$$\mathbb{Q}_p^n \smallsetminus \{0\} = \bigsqcup_{j \in \mathbb{Z}} p^j S_0^n$$

with

$$S_0^n = \{y \in \mathbb{Q}_p^n; \|y\|_p = 1\}.$$

By using this partition and (2.7), we have

$$A_w(\xi) = \sum_{j \in \mathbb{Z}} p^{\gamma n} \int_{p^j S_0^n} \frac{1 - \chi_p(z \cdot \xi_0)}{w(p^\gamma \|z\|_p)} d^n z$$

$$= \sum_{j \in \mathbb{Z}} \frac{p^{-jn+\gamma n}}{w(p^{-j+\gamma})} \left\{ (1 - p^{-n}) - \int_{S_0^n} \chi_p(p^j y \cdot \xi_0) d^n y \right\}.$$

By using the formula

$$\int_{S_0^n} \chi_p(p^j y \cdot \xi_0) d^n y = \begin{cases} 1 - p^{-n} & \text{if } j \geq 0 \\ -p^{-n} & \text{if } j = -1 \\ 0 & \text{if } j < -1, \end{cases} \tag{2.8}$$

see e.g. [105, Lemma 4.1], we get

$$A_w(\xi) = (1 - p^{-n}) \sum_{j=2}^{\infty} \frac{p^{n(\gamma+j)}}{w(p^{\gamma+j})} + \frac{p^{n\gamma+n}}{w(p^{\gamma+1})}$$

$$= (1 - p^{-n}) \sum_{j=\gamma+2}^{\infty} \frac{p^{nj}}{w(p^j)} + \frac{p^{n\gamma+n}}{w(p^{\gamma+1})}. \tag{2.9}$$

From (2.9) follows that $A_w(\xi)$ is radial, positive, continuous outside of the origin, and that $A_w(p^{-ord(\xi)})$ is a decreasing function of $ord(\xi)$. To show that $A_w(p^{-\gamma})$ is a decreasing function of γ, we note that, by (2.9), $A_w(p^{-(\gamma+1)}) - A_w(p^{-\gamma}) = p^{n\gamma+n} \left(\frac{1}{w(p^{\gamma+2})} - \frac{1}{w(p^{\gamma+1})} \right) < 0$. The continuity at the origin follows from

$$A_w(0) := \lim_{\gamma \to \infty} (1 - p^{-n}) \sum_{j=\gamma+2}^{\infty} \frac{p^{nj}}{w(p^j)} + \lim_{\gamma \to \infty} \frac{p^{n\gamma+n}}{w(p^{\gamma+1})} = 0,$$

since $\sum_{j=M}^{\infty} \frac{p^{nj}}{w(p^j)} < \infty$, cf. (2.2), and $\frac{1}{C_0} \geq \frac{p^{\alpha_1(\gamma+1)}}{w(p^{\gamma+1})} \geq \frac{p^{n\gamma+n}}{w(p^{\gamma+1})}$ for γ big enough, cf. (2.1). ∎

Remark 6 We denote by $C(U, \mathbb{C})$, respectively by $C(U, \mathbb{R})$, the vector space of $\mathbb{C}$-valued, respectively of $\mathbb{R}$-valued, continuous functions defined on an open subset U of $\mathbb{Q}_p^n$. In some cases we use the notation $C(U)$, or just C, if there is no danger of confusion.

Proposition 7 *(i)* $(W\varphi)(x) = -\kappa \mathcal{F}_{\xi \to x}^{-1}\left(A_w(\|\xi\|_p)\mathcal{F}_{x \to \xi}\varphi\right)$ *for* $\varphi \in \mathcal{D}\left(\mathbb{Q}_p^n\right)$, *and* $W\varphi \in C\left(\mathbb{Q}_p^n\right) \cap L^\rho\left(\mathbb{Q}_p^n\right)$, *for* $1 \leq \rho \leq \infty$. *The Operator* W *extends to an unbounded and densely defined operator in* $L^2\left(\mathbb{Q}_p^n\right)$ *with domain*

$$Dom(W) = \left\{\varphi \in L^2 ; A_w(\|\xi\|_p)\mathcal{F}\varphi \in L^2\right\}. \tag{2.10}$$

(ii) $(-W, Dom(W))$ *is self-adjoint and positive operator.*
(iii) W *is the infinitesimal generator of a contraction* C_0 *semigroup* $(\mathcal{T}(t))_{t \geq 0}$. *Moreover, the semigroup* $(\mathcal{T}(t))_{t \geq 0}$ *is bounded holomorphic with angle* $\pi/2$.

Proof (i) It follows from Lemma 4 and the fact that $A_w(\|\xi\|_p)$ is continuous, cf. Lemma 5. (ii) follows from the fact that W is a pseudodifferential operator and that the Fourier transform preserves the inner product of L^2. (iii) It follows of well-known results, see e.g. [41, Chap. 2, Sect. 3] or [24]. For the property of the semigroup of being holomorphic, see e.g. [41, Chap. 2, Sect. 4.7]. ∎

2.2.2 Some Additional Results

Lemma 8 *Assume that there exist positive constants* $\alpha_1, \alpha_2, C_0, C_1$, *with* $\alpha_1 > n$, $\alpha_2 > n$, *and* $\alpha_3 \geq 0$, *such that*

$$C_0 \|\xi'\|_p^{\alpha_1} \leq w(\|\xi'\|_p) \leq C_1 \|\xi'\|_p^{\alpha_2} e^{\alpha_3 \|\xi'\|_p}, \text{ for any } \xi' \in \mathbb{Q}_p^n. \tag{2.11}$$

Then there exist positive constants C_2, C_3, *such that*

$$C_2 \|\xi\|_p^{\alpha_2-n} e^{-\alpha_3 p \|\xi\|_p^{-1}} \leq A_w(\|\xi\|_p) \leq C_3 \|\xi\|_p^{\alpha_1-n}$$

for any $\xi \in \mathbb{Q}_p^n$, *with the convention that* $e^{-\alpha_3 p \|0\|_p^{-1}} := \lim_{\|\xi\|_p \to 0} e^{-\alpha_3 p \|\xi\|_p^{-1}} = 0$. *Furthermore, if* $\alpha_3 > 0$, *then* $\alpha_1 \geq \alpha_2$, *and if* $\alpha_3 = 0$, *then* $\alpha_1 = \alpha_2$.

Proof By using the lower bound for w given in (2.11), and $\|\xi\|_p = p^{-\gamma}$,

$$A_w(\|\xi\|_p) \leq \frac{(1-p^{-n})}{C_0} \sum_{j=\gamma+2}^{\infty} \frac{p^{nj}}{p^{j\alpha_1}} + \frac{p^{n\gamma+n}}{p^{\alpha_1(\gamma+1)}} \leq C_3 \|\xi\|_p^{\alpha_1-n}.$$

On the other hand, $A_w\left(\|\xi\|_p\right) \geq \frac{p^{n\gamma+n}}{w(p^{\gamma+1})}$, and by using the upper bound for w given in (2.11),

$$A_w\left(\|\xi\|_p\right) \geq \frac{p^{n\gamma+n}}{w(p^{\gamma+1})} \geq \frac{p^{n\gamma+n}}{C_1 p^{\alpha_2(\gamma+1)} e^{\alpha_3 p^{\gamma+1}}} \geq C_2 \|\xi\|_p^{\alpha_2-n} e^{-\alpha_3 p \|\xi\|_p^{-1}}.$$

∎

Definition 9 We say that W (or A_w) is of exponential type if inequality (2.11) is only possible for $\alpha_3 > 0$ with $\alpha_1, \alpha_2, C_0, C_1$ positive constants and $\alpha_1 > n, \alpha_2 > n$. If (2.11) holds for $\alpha_3 = 0$ with $\alpha_1, \alpha_2, C_0, C_1$ positive constants and $\alpha_1 > n, \alpha_2 > n$, we say that W (or A_w) is of polynomial type.

We note that if W is of polynomial type then $\alpha_1 = \alpha_2 > n$ and C_0, C_1 are positive constants with $C_1 \geq C_0$.

Lemma 10 *With the hypotheses of Lemma 8,*

$$e^{-t\kappa A_w(\|\xi\|_p)} \in L^\rho(\mathbb{Q}_p^n) \text{ for } 1 \leq \rho < \infty \text{ and } t > 0.$$

Proof Since $e^{-tA_w(\|\xi\|_p)}$ is a continuous function, it is sufficient to show that there exists $M \in \mathbb{N}$ such that

$$I_M(t) := \int\limits_{\|\xi\|_p > p^M} e^{-\rho\kappa t A_w(\|\xi\|_p)} d^n\xi < \infty, \text{ for } t > 0.$$

Take $M \in \mathbb{N}$, by Lemma 8, we have

$$C_2 \|\xi\|_p^{\alpha_2-n} e^{-\alpha_3 p \|\xi\|_p^{-1}} > C_2 \|\xi\|_p^{\alpha_2-n} e^{-\alpha_3 p^{-M+1}} \text{ for } \|\xi\|_p > p^M,$$

and (with $B = C_2 \rho\kappa e^{-\alpha_3 p^{-M+1}}$),

$$I_M(t) \leq \int\limits_{\|\xi\|_p > p^M} e^{-tB\|\xi\|_p^{\alpha_2-n}} d^n\xi \leq C(M,\kappa,\rho) t^{\frac{-n}{\alpha_2-n}}, \text{ for } t > 0.$$

∎

2.2.3 *p-Adic Description of Characteristic Relaxation in Complex Systems*

In [11] Avetisov et al. developed a new approach to the description of relaxation processes in complex systems (such as glasses, macromolecules and proteins) on the basis of *p*-adic analysis. The dynamics of a complex system is described by a random walk in the space of configurational states, which is approximated by an ultrametric space ($\mathbb{Q}_p$). Mathematically speaking, the time-evolution of the system is controlled by a master equation of the form

$$\frac{\partial f(x,t)}{\partial t} = \int_{\mathbb{Q}_p} \{v(x \mid y)f(y,t) - v(y \mid x)f(x,t)\}\,dy, \ x \in \mathbb{Q}_p, \ t \in \mathbb{R}_+, \qquad (2.12)$$

where the function $f(x,t) : \mathbb{Q}_p \times \mathbb{R}_+ \to \mathbb{R}_+$ is a probability density distribution, and the function $v(x \mid y) : \mathbb{Q}_p \times \mathbb{Q}_p \to \mathbb{R}_+$ is the probability of transition from state y to the state x per unit time. The transition from a state y to a state x can be perceived as overcoming the energy barrier separating these states. In [11] an Arrhenius type relation was used:

$$v(x \mid y) \sim A(T) \exp\left\{-\frac{U(x \mid y)}{kT}\right\},$$

where $U(x \mid y)$ is the height of the activation barrier for the transition from the state y to state x, k is the Boltzmann constant and T is the temperature. This formula establishes a relation between the structure of *the energy landscape $U(x \mid y)$* and the transition function $v(x \mid y)$. The case $v(x \mid y) = v(y \mid x)$ corresponds to a *degenerate energy landscape*. In this case the master equation (2.12) takes the form

$$\frac{\partial f(x,t)}{\partial t} = \int_{\mathbb{Q}_p} v\left(|x - y|_p\right) \{f(y,t) - f(x,t)\}\,dy,$$

where $v\left(|x - y|_p\right) = \frac{A(T)}{|x-y|_p} \exp\left\{-\frac{U(|x-y|_p)}{kT}\right\}$. By choosing U conveniently, several energy landscapes can be obtained. Following [11], there are three basic landscapes: (i) (logarithmic) $v\left(|x - y|_p\right) = \frac{1}{|x-y|_p \ln^\alpha \left(1+|x-y|_p\right)}, \alpha > 1$; (ii) (linear) $v\left(|x - y|_p\right) = \frac{1}{|x-y|_p^{\alpha+1}}, \alpha > 0$; (iii) (exponential) $v\left(|x - y|_p\right) = \frac{e^{-\alpha|x-y|_p}}{|x-y|_p}, \alpha > 0$.

Thus, it is natural to study the following Cauchy problem:

$$\begin{cases} \frac{\partial u(x,t)}{\partial t} = \kappa \int_{\mathbb{Q}_p^n} \frac{u(x-y,t)-u(x,t)}{w(y)}\,d^n y, \ x \in \mathbb{Q}_p^n, \ t \in \mathbb{R}_+, \\[3mm] u(x,0) = \varphi \in \mathcal{D}\left(\mathbb{Q}_p^n\right), \end{cases}$$

where $w(y)$ is a radial function belonging to a class of functions that contains functions like:

(i) $w(\|y\|_p) = \Gamma_p^n(-\alpha) \|y\|_p^{\alpha+n}$, here $\Gamma_p^n(\cdot)$ is the n-dimensional p-adic Gamma function, and $\alpha > 0$;
(ii) $w(\|y\|_p) = \|y\|_p^\beta e^{\alpha\|y\|_p}, \alpha > 0$.

By imposing condition (2.11) to w, we include the linear and exponential energy landscapes in our study. On the other hand, take $w(\|y\|_p)$ satisfying (2.11) and take $f\left(\|y\|_p\right)$ a continuous and increasing function such that

$$0 < \sup_{y \in \mathbb{Q}_p^n} f\left(\|y\|_p\right) < \infty \text{ and } 0 < \inf_{y \in \mathbb{Q}_p^n} f\left(\|y\|_p\right) < \infty.$$

Then $f\left(\|y\|_p\right) w(\|y\|_p)$ satisfies (2.11). This fact shows that the class of operators $\boldsymbol{W}$ is very large.

Finally we note that $\|y\|_p^\beta \ln^\alpha (1 + \|y\|_p)$, $\beta > n$, $\alpha \in \mathbb{N}$, does not satisfies $\|y\|_p^{\alpha_1} \le \|y\|_p^\beta \ln^\alpha (1 + \|y\|_p)$ for any $y \in \mathbb{Q}_p^n$, and hence our results do not include the case of logarithmic landscapes.

2.2.4 Heat Kernels

In this section we assume that function w satisfies conditions (2.11). We define

$$Z(x, t; w, \kappa) := Z(x, t) = \int_{\mathbb{Q}_p^n} e^{-\kappa t A_w(\|\xi\|_p)} \chi_p(-x \cdot \xi) d^n \xi \text{ for } t > 0 \text{ and } x \in \mathbb{Q}_p^n.$$

Note that by Lemma 10, $Z(x, t) = \mathcal{F}_{\xi \to x}^{-1}[e^{-\kappa t A_w(\|\xi\|_p)}] \in C \cap L^2$ for $t > 0$. We call a such function a *heat kernel*. When considering $Z(x, t)$ as a function of x for t fixed we will write $Z_t(x)$.

Lemma 11 *(i) There exists a positive constant C, such that*

$$Z(x, t) < Ct \|x\|_p^{-\alpha_1}, \text{ for } x \in \mathbb{Q}_p^n \smallsetminus \{0\} \text{ and } t > 0.$$

(ii) $Z_t(x) \in L^1\left(\mathbb{Q}_p^n\right)$ for every $t > 0$.

Proof (i) Let $\|x\|_p = p^\beta$. Since $Z(x, t) \in L^1(\mathbb{Q}_p^n)$ for $t > 0$, by using $\mathbb{Q}_p^n \smallsetminus \{0\} = \bigsqcup_{j \in \mathbb{Z}} p^j S_0^n$ and formula (2.8), we get

$$Z(x, t) = \|x\|_p^{-n} \left[(1 - p^{-n}) \sum_{j=0}^{\infty} e^{-\kappa A_w(p^{-\beta-j})t} p^{-nj} - e^{-\kappa A_w(p^{-\beta+1})t} \right].$$

By using that $e^{-\kappa A_w(p^{-\beta-j})t} \le 1$ for $j \in \mathbb{N}$, we have

$$Z(x, t) \le \|x\|_p^{-n} \left[1 - e^{-\kappa A_w(p^{-\beta+1})t} \right].$$

We now apply the mean value theorem to the real function $f(u) = e^{-\kappa A_w(p^{-\beta+1})u}$ on $[0, t]$ with $t > 0$, and Lemma 8,

$$Z(x, t) \leq C_0 \|x\|_p^{-n} t A_w(p^{-\beta+1}) \leq Ct \|x\|_p^{-\alpha_1}.$$

(ii) Notice that

$$\int_{\mathbb{Q}_p^n} Z_t(x) d^n x = \int_{B_0^n} Z_t(x) d^n x + \int_{\mathbb{Q}_p^n \setminus B_0^n} Z_t(x) d^n x,$$

the existence of the first integral follows from the continuity of $Z_t(x)$, for the second integral we use the bound obtained in (i).

∎

Lemma 12 $Z(x, t) \geq 0$, for $x \in \mathbb{Q}_p^n$ and $t > 0$.

Proof Since $e^{-\kappa t A_w(\|\xi\|_p)}$ is radial, by using $\mathbb{Q}_p^n \setminus \{0\} = \bigsqcup_{j \in \mathbb{Z}} p^j S_0^n$ and formula (2.8), we have

$$Z(x, t) = \sum_{i=-\infty}^{\infty} e^{-\kappa t A_w(p^i)} \int_{\|\xi\|_p = p^i} \chi_p(-x \cdot \xi) d^n \xi$$

$$= \sum_{i=-\infty}^{\infty} p^{ni} \left[e^{-\kappa t A_w(p^i)} - e^{-\kappa t A_w(p^{i+1})} \right] \Omega(\|p^{-i} x\|_p) \geq 0$$

since A_w is increasing function of i, cf. Lemma 5.

∎

Theorem 13 *The function $Z(x, t)$ has the following properties:*

(i) $Z(x, t) \geq 0$ for any $t > 0$;
(ii) $\int_{\mathbb{Q}_p^n} Z(x, t) d^n x = 1$ for any $t > 0$;
(iii) $Z_t(x) \in C(\mathbb{Q}_p^n, \mathbb{R}) \cap L^1(\mathbb{Q}_p^n) \cap L^2(\mathbb{Q}_p^n)$ for any $t > 0$;
(iv) $Z_t(x) * Z_{t'}(x) = Z_{t+t'}(x)$ for any $t, t' > 0$;
(v) $\lim_{t \to 0^+} Z(x, t) = \delta(x)$ in $\mathcal{D}'(\mathbb{Q}_p^n)$, where δ denotes the Dirac distribution.

Proof (i) It follows from Lemma 12. (ii) Since $Z_t(x)$, $\mathcal{F}_{x \to \xi}(Z_t(x)) = e^{-\kappa t A_w(\|\xi\|_p)} \in C \cap L^1$, for any $t > 0$, cf. Lemma 10 and Lemma 11 (ii), the result follows from the inversion formula for the Fourier transform. (iii) It follows from Lemma 10 and Lemma 11 (ii). (iv) By the previous property $Z_t(x) \in L^1$ for any $t > 0$, then

$$Z_t(x) * Z_{t'}(x) = \mathcal{F}_{\xi \to x}^{-1} \left(e^{-\kappa t A_w(\|\xi\|_p)} e^{-\kappa t' A_w(\|\xi\|_p)} \right)$$

$$= \mathcal{F}_{\xi \to x}^{-1} \left(e^{-\kappa(t+t') A_w(\|\xi\|_p)} \right) = Z_{t+t'}(x).$$

(v) Since we have $e^{-\kappa t A_w(\|\xi\|_p)} \in C(\mathbb{Q}_p^n, \mathbb{R}) \cap L^1$ for $t > 0$, cf. Lemma 10, the inner product

$$\left\langle e^{-\kappa t A_w(\|\xi\|_p)}, \phi \right\rangle = \int_{\mathbb{Q}_p^n} e^{-\kappa t A_w(\|\xi\|_p)} \overline{\phi(\xi)} d^n\xi$$

defines a distribution on $\mathbb{Q}_p^n$, then, by the dominated convergence theorem,

$$\lim_{t \to 0^+} \left\langle e^{-\kappa t A_w(\|\xi\|_p)}, \phi \right\rangle = \langle 1, \phi \rangle$$

and thus

$$\lim_{t \to 0^+} \langle Z(x, t), \phi \rangle = \lim_{t \to 0^+} \left\langle e^{-\kappa t A_w(\|\xi\|_p)}, \mathcal{F}^{-1}\phi \right\rangle = \left\langle 1, \mathcal{F}^{-1}\phi \right\rangle = (\delta, \phi).$$

∎

2.2.5 Markov Processes Over $\mathbb{Q}_p^n$

Along this section we consider $(\mathbb{Q}_p^n, \|\cdot\|_p)$ as complete non-Archimedean metric space and use the terminology and results of [39, Chapters 2, 3]. Let $\mathcal{B}$ denote the Borel σ-algebra of $\mathbb{Q}_p^n$. Thus $(\mathbb{Q}_p^n, \mathcal{B}, d^n x)$ is a measure space.

We set

$$p(t, x, y) := Z(x - y, t) \text{ for } t > 0, \, x, y \in \mathbb{Q}_p^n,$$

and

$$P(t, x, B) = \begin{cases} \int_B p(t, y, x) d^n y & \text{for } t > 0, \quad x \in \mathbb{Q}_p^n, \quad B \in \mathcal{B} \\ 1_B(x) & \text{for } t = 0. \end{cases}$$

Lemma 14 *With the above notation the following assertions hold:*

(i) $p(t, x, y)$ is a normal transition density;
(ii) $P(t, x, B)$ is a normal transition function.

Proof The result follows from Theorem 13, see [39, Section 2.1] for further details. ∎

Lemma 15 *The transition function $P(t, x, B)$ satisfies the following two conditions:*

(i) for each $u \geq 0$ and compact B

$$\lim_{x \to \infty} \sup_{t \leq u} P(t, x, B) = 0 \ [Condition \ L(B)];$$

(ii) for each $\epsilon > 0$ and compact B

$$\lim_{t \to 0^+} \sup_{x \in B} P(t, x, \mathbb{Q}_p^n \setminus B_\epsilon^n(x)) = 0 \ [Condition \ M(B)].$$

Proof (i) By Lemma 11 and the fact that $\|\cdot\|_p$ is an ultranorm, we have

$$P(t, x, B) \leq Ct \int_B \|x - y\|_p^{-\alpha_1} d^n y = tC \|x\|_p^{-\alpha_1} vol(B) \text{ for } x \in \mathbb{Q}_p^n \setminus B.$$

Therefore $\lim_{x \to \infty} \sup_{t \leq u} P(t, x, B) = 0$.

(ii) Again, by Lemma 11, the fact that $\|\cdot\|_p$ is an ultranorm, and $\alpha_1 > n$, we have

$$P(t, x, \mathbb{Q}_p^n \setminus B_\epsilon^n(x)) \leq Ct \int_{\|x-y\|_p > \epsilon} \|x - y\|_p^{-\alpha_1} d^n y = Ct \int_{\|z\|_p > \epsilon} \|z\|_p^{-\alpha_1} d^n z$$

$$= C'(\alpha_1, \epsilon, n) t.$$

Therefore

$$\lim_{t \to 0^+} \sup_{x \in B} P(t, x, \mathbb{Q}_p^n \setminus B_\epsilon^n(x)) \leq \lim_{t \to 0^+} \sup_{x \in B} C'(\alpha_1, \epsilon, n) t = 0.$$

∎

Theorem 16 $Z(x, t)$ *is the transition density of a time and space homogeneous Markov process which is bounded, right-continuous and has no discontinuities other than jumps.*

Proof The result follows from [39, Theorem 3.6] by using that $(\mathbb{Q}_p^n, \|x\|_p)$ is semicompact space, i.e. a locally compact Hausdorff space with a countable base, and $P(t, x, B)$ is a normal transition function satisfying conditions $L(B)$ and $M(B)$, cf. Lemmas 14, 15.

∎

2.2.6 The Cauchy Problem

Consider the following Cauchy problem:

$$
\begin{cases}
\frac{\partial u}{\partial t}(x,t) - \boldsymbol{W}u(x,t) = 0, \ x \in \mathbb{Q}_p^n, t \in [0,\infty), \\[2ex]
u(x,0) = u_0(x), \qquad\qquad u_0(x) \in Dom(\boldsymbol{W}),
\end{cases}
\tag{2.13}
$$

where $(\boldsymbol{W}\phi)(x) = -\kappa \mathcal{F}_{\xi \to x}^{-1}\left(A_w\left(\|\xi\|_p\right)\mathcal{F}_{x \to \xi}\phi\right)$ for $\phi \in Dom(\boldsymbol{W})$, see (2.10), and $u : \mathbb{Q}_p^n \times [0,\infty) \to \mathbb{C}$ is an unknown function. We say that a function $u(x,t)$ is a *solution* of (2.13), if $u(x,t) \in C\left([0,\infty), Dom(\boldsymbol{W})\right) \cap C^1\left([0,\infty), L^2(\mathbb{Q}_p^n)\right)$ and u satisfies (2.13) for all $t \geq 0$.

In this section, we understand the notions of continuity in t, differentiability in t and equalities in the $L^2(\mathbb{Q}_p^n)$ sense, as it is customary in the semigroup theory.

We know from Proposition 7 that the operator $\boldsymbol{W}$ generates a C_0 semigroup $(\mathcal{T}(t))_{t \geq 0}$, then Cauchy problem (2.13) is well-posed, i.e. it is uniquely solvable with the solution continuously dependent on the initial datum, and its solution is given by $u(x,t) = \mathcal{T}(t)u_0(x)$, for $t \geq 0$, see e.g. [24, Theorem 3.1.1]. However the general theory does not give an explicit formula for the semigroup $(\mathcal{T}(t))_{t \geq 0}$. We show that the operator $\mathcal{T}(t)$ for $t > 0$ coincides with the operator of convolution with the heat kernel $Z_t * \cdot$. In order to prove this, we first construct a solution of Cauchy problem (2.13) with the initial value from $\mathcal{D}$ without using the semigroup theory. Then we extend the result to all initial values from $Dom(\boldsymbol{W})$, see Propositions 18–20.

2.2.6.1 Homogeneous Equations with Initial Values in $\mathcal{D}$

To simplify the notation, set $Z_0 * u_0 = (Z_t(x) * u_0(x))\,|_{t=0} := u_0$. We define the function

$$
u(x,t) = Z_t(x) * u_0(x), \text{ for } t \geq 0. \tag{2.14}
$$

Since $Z_t(x) \in L^1$ for $t > 0$ and $u_0 \in \mathcal{D}(\mathbb{Q}_p^n) \subset L^\infty(\mathbb{Q}_p^n)$, the convolution exists and is a continuous function, see e.g. [100, Theorem 1.1.6].

Lemma 17 *Take $u_0 \in \mathcal{D}$ with the support of $\widehat{u_0}$ contained in B_R^n, and $u(x,t)$, $t \geq 0$ defined as in (2.14). Then the following assertions hold:*

(i) *$u(x,t)$ is continuously differentiable in time for $t \geq 0$ and the derivative is given by*

$$
\frac{\partial u(x,t)}{\partial t} = -\kappa \mathcal{F}_{\xi \to x}^{-1}\left(e^{-\kappa t A_w(\|\xi\|_p)}A_w(\|\xi\|_p)1_{B_R^n}(\xi)\right) * u_0(x);
$$

(ii) $u(x,t) \in Dom(W)$ *for any* $t \geq 0$ *and*

$$(Wu)(x,t) = -\kappa \mathcal{F}^{-1}_{\xi \to x} \left(e^{-\kappa t A_w(\|\xi\|_p)} A_w(\|\xi\|_p) 1_{B^n_R}(\xi) \right) * u_0(x).$$

Proof (i) The proof is similar to the one given for Lemma 110 in Chap. 4. (ii) Note that $e^{-\kappa t A_w(\|\xi\|_p)} \widehat{u_0}(\xi)$, $A_w(\|\xi\|_p) e^{-\kappa t A_w(\|\xi\|_p)} \widehat{u_0}(\xi) \in C \cap L^2 \cap L^1$ for $t \geq 0$, i.e. $u(x,t) \in Dom(W)$ for $t \geq 0$. Now

$$(Wu)(x,t) = -\kappa \mathcal{F}^{-1}_{\xi \to x} \left(A_w(\|\xi\|_p) \mathcal{F}_{\xi \to x}(u(x,t)) \right)$$

$$= -\kappa \mathcal{F}^{-1}_{\xi \to x} \left(A_w(\|\xi\|_p) e^{-\kappa t A_w(\|\xi\|_p)} \widehat{u_0}(\xi) \right)$$

$$= -\kappa \mathcal{F}^{-1}_{\xi \to x} \left(A_w(\|\xi\|_p) e^{-\kappa t A_w(\|\xi\|_p)} 1_{B^n_R}(\xi) \widehat{u_0}(\xi) \right)$$

$$= -\kappa \mathcal{F}^{-1}_{\xi \to x} \left(e^{-\kappa t A_w(\|\xi\|_p)} A_w(\|\xi\|_p) 1_{B^n_R}(\xi) \right) * u_0(x).$$

∎

As a direct consequence of Lemma 17 we obtain the following result.

Proposition 18 *Assume that $u_0 \in \mathcal{D}$. Then function $u(x,t)$ defined in (2.14) is a solution of Cauchy problem (2.13).*

2.2.6.2 Homogeneous Equations with Initial Values in L^2

We define

$$T(t)u = \begin{cases} Z_t * u, & t > 0 \\ \\ u, & t = 0, \end{cases} \tag{2.15}$$

for $u \in L^2$.

Lemma 19 *The operator $T(t) : L^2(\mathbb{Q}^n_p) \longrightarrow L^2(\mathbb{Q}^n_p)$ is bounded for any fixed $t \geq 0$.*

Proof For $t > 0$, the result follows from the Young inequality by using the fact that $Z_t \in L^1$, cf. Theorem 13 (iii). ∎

Proposition 20 *The following assertions hold.*

(i) *The operator W generates a C_0 semigroup $(T(t))_{t \geq 0}$. The operator $T(t)$ coincides for each $t \geq 0$ with the operator $T(t)$ given by (2.15).*
(ii) *Cauchy problem (2.13) is well-posed and its solution is given by $u(x,t) = Z_t * u_0$, $t \geq 0$.*

Proof (i) By Proposition 7 (iii) the operator W generates a C_0 semigroup $(\mathcal{T}(t))_{t\geq0}$. Hence Cauchy problem (2.13) is well-posed, see e.g. [24, Theorem 3.1.1]. By Proposition 18, $\mathcal{T}(t)|_{\mathcal{D}} = T(t)|_{\mathcal{D}}$ and both operators $\mathcal{T}(t)$ and $T(t)$ are defined on the whole L^2 and bounded, cf. Lemma 19. By the continuity we conclude that $\mathcal{T}(t)| = T(t)$ on L^2. Now the statements follow from well-known results of the semigroup theory, see e.g. [24, Theorem 3.1.1], [41, Chap. 2, Proposition 6.2]. ∎

2.2.6.3 Non-homogeneous Equations

Consider the following Cauchy problem:

$$\begin{cases} \frac{\partial u}{\partial t}(x,t) - Wu(x,t) = g(x,t),\ x \in \mathbb{Q}_p^n, t \in [0,T]\,, T > 0, \\[2mm] u(x,0) = u_0(x), \qquad\qquad u_0(x) \in Dom(W). \end{cases} \qquad (2.16)$$

We say that a function $u(x,t)$ is a solution of (2.16), if $u(x,t)$ belongs to $C\left([0,T), Dom(W)\right) \cap C^1\left([0,T], L^2(\mathbb{Q}_p^n)\right)$ and if $u(x,t)$ satisfies equation (2.16) for $t \in [0,T]$.

Theorem 21 *Assume that* $u_0 \in Dom(W)$ *and* $g \in C\left([0,\infty), L^2(\mathbb{Q}_p^n)\right) \cap L^1\left((0,\infty), Dom(W)\right)$. *Then Cauchy problem (2.16) has a unique solution given by*

$$u(x,t) = \int_{\mathbb{Q}_p^n} Z(x-\xi,t)u_0(\xi)d^n\xi + \int_0^t \int_{\mathbb{Q}_p^n} Z(x-\xi,t-\theta)g(\xi,\theta)d^n\xi d\theta.$$

Proof The result follows from Proposition 20 by using some well-known results of the semigroup theory, see e.g. [24, Proposition 4.1.6]. ∎

2.2.7 The Taibleson Operator and the p-Adic Heat Equation

We set

$$\Gamma_p^{(n)}(\alpha) := \frac{1-p^{\alpha-n}}{1-p^{-\alpha}}, \text{ for } \alpha \in \mathbb{R} \setminus \{0\}\,.$$

This function is called the *p-adic Gamma function*. The function

$$k_\alpha(x) = \frac{||x||_p^{\alpha-n}}{\Gamma_p^{(n)}(\alpha)}, \quad \alpha \in \mathbb{R} \setminus \{0,n\}\,, \quad x \in \mathbb{Q}_p^n,$$

is called *the multi-dimensional Riesz Kernel*; it determines a distribution on $\mathcal{D}(\mathbb{Q}_p^n)$ as follows. If $\alpha \neq 0, n$, and $\varphi \in \mathcal{D}(\mathbb{Q}_p^n)$, then

$$(k_\alpha(x), \varphi(x)) = \frac{1 - p^{-n}}{1 - p^{\alpha - n}} \varphi(0) + \frac{1 - p^{-\alpha}}{1 - p^{\alpha - n}} \int_{||x||_p > 1} ||x||_p^{\alpha - n} \varphi(x) \, d^n x$$

$$+ \frac{1 - p^{-\alpha}}{1 - p^{\alpha - n}} \int_{||x||_p \leq 1} ||x||_p^{\alpha - n} (\varphi(x) - \varphi(0)) \, d^n x. \qquad (2.17)$$

Then $k_\alpha \in \mathcal{D}'(\mathbb{Q}_p^n)$, for $\mathbb{R} \setminus \{0, n\}$. In the case $\alpha = 0$, by passing to the limit in (2.17), we obtain

$$(k_0(x), \varphi(x)) := \lim_{\alpha \to 0} (k_\alpha(x), \varphi(x)) = \varphi(0),$$

i.e., $k_0(x) = \delta(x)$, the Dirac delta function, and therefore $k_\alpha \in \mathcal{D}'(\mathbb{Q}_p^n)$, for $\mathbb{R} \setminus \{n\}$.

It follows from (2.17) that for $\alpha > 0$,

$$(k_{-\alpha}(x), \varphi(x)) = \frac{1 - p^\alpha}{1 - p^{-\alpha - n}} \int_{\mathbb{Q}_p^n} ||x||_p^{-\alpha - n} (\varphi(x) - \varphi(0)) \, d^n x. \qquad (2.18)$$

Definition 22   The Taibleson pseudodifferential operator D_T^α, $\alpha > 0$, is defined as

$$(D_T^\alpha \varphi)(x) = \mathcal{F}_{\xi \to x}^{-1} \left(||\xi||_p^\alpha \mathcal{F}_{x \to \xi} \varphi \right), \text{ for } \varphi \in \mathcal{D}(\mathbb{Q}_p^n).$$

By using (2.18) and the fact that $(\mathcal{F} k_{-\alpha})(x)$ equals $||x||_p^\alpha$, $\alpha \neq -n$, in $\mathcal{D}'(\mathbb{Q}_p^n)$, we have

$$\left(D_T^\alpha \varphi \right)(x) = (k_{-\alpha} * \varphi)(x)$$

$$= \frac{1 - p^\alpha}{1 - p^{-\alpha - n}} \int_{\mathbb{Q}_p^n} ||y||_p^{-\alpha - n} (\varphi(x - y) - \varphi(x)) \, d^n y. \qquad (2.19)$$

Then the Taibleson operator belongs to the class of operators *W* introduced before. The right-hand side of (2.19) makes sense for a wider class of functions, for example, for locally constant functions $\varphi(x)$ satisfying

$$\int_{||x||_p \geq 1} ||x||_p^{-\alpha - n} |\varphi(x)| \, d^n x < \infty.$$

A similar observation is valid in general for operators of *W* type. The equation

$$\frac{\partial u(x, t)}{\partial t} + \kappa (D_T^\alpha u)(x, t) = 0, \quad x \in \mathbb{Q}_p^n, \quad t \geq 0,$$

where κ is a positive constant, is a multi-dimensional analog of the p-adic heat equation introduced in [111].

2.3 Elliptic Pseudodifferential Operators, Parabolic-Type Equations and Markov Processes

In this section we consider following Cauchy problem:

$$\begin{cases} \frac{\partial u(x,t)}{\partial t} + (f(\partial, \beta) u)(x,t) = 0, \ x \in \mathbb{Q}_p^n, n \geq 1, t \geq 0 \\ \\ u(x,0) = \varphi(x), \end{cases} \tag{2.20}$$

where $f(\partial, \beta)$ is an elliptic pseudodifferential operator of the form

$$(f(\partial, \beta) \phi)(x,t) = \mathcal{F}_{\xi \to x}^{-1}\left(|f(\xi)|_p^{\beta} \mathcal{F}_{x \to \xi} \phi(x,t)\right).$$

Here β is a positive real number, and $f(\xi) \in \mathbb{Q}_p[\xi_1, \ldots, \xi_n]$ is a homogeneous polynomial of degree d satisfying the property $f(\xi) = 0 \Leftrightarrow \xi = 0$. We establish the existence of a unique solution to Cuachy problem (2.20) in the case in which $\varphi(x)$ is a continuous and an integrable function. Under these hypotheses we show the existence of a solution $u(x,t)$ that is continuous in x, for a fixed $t \in [0, T]$, bounded, and integrable function. In addition the solution can be presented in the form

$$u(x,t) = Z(x,t) * \varphi(x)$$

where $Z(x,t)$ is *the fundamental solution* (also called *the heat kernel*) to Cauchy's Problem 2.20:

$$Z(x,t,f,\beta) := Z(x,t) = \int_{\mathbb{Q}_p^n} \chi_p(-x \cdot \xi) e^{-t|f(\xi)|_p^{\beta}} d^n\xi, \ \xi \in \mathbb{Q}_p^n, t > 0. \tag{2.21}$$

The fundamental solution is a transition density of a Markov process with space state $\mathbb{Q}_p^n$.

2.3.1 Elliptic Operators

Let $h(\xi) \in \mathbb{Q}_p[\xi_1, \ldots, \xi_n]$ be a non-constant polynomial. In this section we work with operators of the form $h(\partial, \beta) \phi = \mathcal{F}^{-1}\left(|h|_p^{\beta} \mathcal{F}\phi\right)$, $\beta > 0$, $\phi \in \mathcal{D}(\mathbb{Q}_p^n)$. We will say that $h(\partial, \beta)$ is *a pseudodifferential operator with symbol* $|h|_p^\beta = |h(\xi)|_p^\beta$. The operator $h(\partial, \beta)$ has a self-adjoint extension with dense domain in L^2.

Definition 23 Let $f(\xi) \in \mathbb{Q}_p[\xi_1, \ldots, \xi_n]$ be a non-constant polynomial. We say that $f(\xi)$ is an elliptic polynomial of degree d, if it satisfies: (i) $f(\xi)$ is a homogeneous polynomial of degree d, and (ii) $f(\xi) = 0 \Leftrightarrow \xi = 0$.

Lemma 24 *(i) There are infinitely many elliptic polynomials. (ii) For any $n \in \mathbb{N} \setminus \{0\}$ and $p \neq 2$, there exists an elliptic polynomial $h(\xi_1, \ldots, \xi_n)$ with coefficients in $\mathbb{Z}_p^\times$ and degree $2d(n) := 2d$ such that*

$$|h(\xi_1, \ldots, \xi_n)|_p = \|(\xi_1, \ldots, \xi_n)\|_p^{2d}. \tag{2.22}$$

Proof (i) Assume that $h(\xi_1, \ldots, \xi_n)$ is an elliptic polynomial of degree d. Take $\tau \in \mathbb{Q}_p^\times$ such that the equation $x^2 = \tau$ has no solutions in $\mathbb{Q}_p^\times$, then $h(\xi_1, \ldots, \xi_n)^2 - \tau \xi_{n+1}^{2d}$ is an elliptic polynomial of degree $2d$. Since there are elliptic quadratic forms for $1 \leq n \leq 4$, see e.g. [22, Chapter 1], one concludes the existence of infinitely many elliptic polynomials. (ii) By choosing $\tau \in \mathbb{Z}_p^\times$, it follows from (i) that if $h(\xi_1, \ldots, \xi_n)$ is an elliptic polynomials of degree d with coefficients in $\mathbb{Z}_p^\times$, then $h(\xi_1, \ldots, \xi_n)^2 - \tau \xi_{n+1}^{2d}$ is elliptic with coefficients in $\mathbb{Z}_p^\times$. We pick d such that p does not divide d. We prove by induction on n that $h(\xi_1, \ldots, \xi_n)^2 - \tau \xi_{n+1}^{2d}$ satisfies (2.22). Assume, as induction hypothesis, that $h(\xi_1, \ldots, \xi_n)$ satisfies (2.22). If $\left| h(\xi_1, \ldots, \xi_n)^2 \right|_p \neq \left| \xi_{n+1}^{2d} \right|_p$, then $\left| h(\xi_1, \ldots, \xi_n)^2 - \tau \xi_{n+1}^{2d} \right|_p = \left\| (\xi_1, \ldots, \xi_{n+1}) \right\|_p^{2d}$. If $\left| h(\xi_1, \ldots, \xi_n)^2 \right|_p = \left| \xi_{n+1}^{2d} \right|_p$, taking $\xi_{n+1} = p^m u_{n+1}$, with $u_{n+1} \in \mathbb{Z}_p^\times$, we have

$$\left| h(\xi_1, \ldots, \xi_n)^2 - \tau \xi_{n+1}^{2d} \right|_p = p^{-2md} \left| h(p^{-m}\xi_1, \ldots, p^{-m}\xi_n)^2 - \tau u_{n+1}^{2d} \right|_p.$$

We note that $h(p^{-1}\xi_1, \ldots, p^{-1}\xi_n)^2 - \tau u_{n+1}^{2d} \in \mathbb{Z}_p^\times$, otherwise

$$h(p^{-m}\xi_1, \ldots, p^{-m}\xi_n)^2 - \tau u_{n+1}^{2d} \equiv 0 \bmod p$$

and by using that p does not divide $2d$, i.e. $p \neq 2$ and the Hensel lemma, there exists a nontrivial solution of $h(\xi_1, \ldots, \xi_n)^2 - \tau \xi_{n+1}^{2d} = 0$, which is impossible. Finally, by using $\left| h(\xi_1, \ldots, \xi_n)^2 \right|_p = \left\| (\xi_1, \ldots, \xi_n) \right\|_p^{2d} = \left| \xi_{n+1} \right|_p^{2d} = p^{-2md}$, we have $\left| h(\xi_1, \ldots, \xi_n)^2 - \tau \xi_{n+1}^{2d} \right|_p = \left\| (\xi_1, \ldots, \xi_{n+1}) \right\|_p^{2d}$. ∎

Lemma 25 *Let $f(\xi) \in \mathbb{Q}_p[\xi]$, $\xi = (\xi_1, \ldots, \xi_n)$, be an elliptic polynomial of degree d. Then there exist positive constants $C_0 = C_0(f)$, $C_1 = C_1(f)$ such that*

$$C_0 \|\xi\|_p^d \leq |f(\xi)|_p \leq C_1 \|\xi\|_p^d, \text{ for every } \xi \in \mathbb{Q}_p^n. \tag{2.23}$$

Proof Without loss of generality we may assume that $\xi \neq 0$. Let $\tilde{\xi} \in \mathbb{Q}_p^{\times}$ be an element such that $\left| \tilde{\xi} \right|_p = \|\xi\|_p \neq 0$. We first note that

$$|f(\xi)|_p = \left| \tilde{\xi} \right|_p^d \left| f\left(\tilde{\xi}^{-1} \xi \right) \right|_p, \tag{2.24}$$

with $\tilde{\xi}^{-1} \xi \in S_0^n = \{z \in \mathbb{Z}_p^n; \|z\|_p = 1\}$. Now $|f|_p$ is continuous on S_0^n, that is a compact subset of $\mathbb{Z}_p^n$, then $\inf_{z \in S_0^n} |f(z)|_p$, and $\sup_{z \in S_0^n} |f(z)|_p$ are attained on S_0^n, and since $|f|_p > 0$ on S_0^n, we have $\sup_{z \in S_0^n} |f(z)|_p \geq \inf_{z \in S_0^n} |f(z)|_p > 0$. Therefore from (2.24) we have

$$\left(\inf_{z \in S_0^n} |f(z)|_p \right) \left| \tilde{\xi} \right|_p^d \leq |f(\xi)|_p \leq \left(\sup_{z \in S_0^n} |f(z)|_p \right) \left| \tilde{\xi} \right|_p^d.$$

∎

Along this section $f(\xi)$ will denote an elliptic polynomial of degree d. Now, since $cf(\xi)$ is elliptic for any $c \in \mathbb{Q}_p^{\times}$ when $f(\xi)$ is elliptic, we will assume that all the elliptic polynomials have coefficients in $\mathbb{Z}_p$.

Lemma 26 *Let $A \subseteq \mathbb{Q}_p^n$ be an open compact subset such that $0 \notin A$. There exist a finite number of points $\tilde{\xi}_i \in A$, $i = 1, \cdots, L_0$, and a constant $M := M(A, f) \in \mathbb{N} \smallsetminus \{0\}$ such that*

$$A = \bigsqcup_{i=1}^{L_0} \tilde{\xi}_i + \left(p^M \mathbb{Z}_p \right)^n \text{ and } |f(\xi)|_p \mid_{\tilde{\xi}_i + (p^M \mathbb{Z}_p)^n} = \left| f\left(\tilde{\xi}_i \right) \right|_p, i = 1, \cdots, L_0.$$

Proof By (2.23), for $\xi \in A$,

$$|f(\xi)|_p \geq C_0 \|\xi\|_p^d \geq C_0 \inf_{\xi \in A} \|\xi\|_p^d \geq p^{-M'(A,f)},$$

where $M' := M'(A, f)$ is a positive integer constant. Now for $\tilde{\xi}_i \in A$ and $y \in \mathbb{Z}_p^n$,

$$f\left(\tilde{\xi}_i + p^M y \right) = f\left(\tilde{\xi}_i \right) + p^M T\left(\tilde{\xi}_i, y \right),$$

where $T\left(\tilde{\xi}_i, y \right)$ is a polynomial function in $\tilde{\xi}_i$, y, with

$$\sup_{\tilde{\xi}_i \in A, y \in \mathbb{Z}_p^n} \left| T\left(\tilde{\xi}_i, y \right) \right|_p \leq p^{\delta}.$$

We set $M = M' + \delta + 1$. Then

$$\left| f\left(\tilde{\xi}_i + p^M y\right)\right|_p = \left| f\left(\tilde{\xi}_i\right) + p^M T\left(\tilde{\xi}_i, y\right)\right|_p = \left| f\left(\tilde{\xi}_i\right)\right|_p .$$

Now, since A is open compact, there exist a finite number of points $\tilde{\xi}_i \in A$, such that $A = \bigsqcup_i \tilde{\xi}_i + \left(p^M \mathbb{Z}_p\right)^n$. ∎

Remark 27 Lemma 26 is valid for arbitrary polynomials satisfying only $f(\xi) = 0 \Leftrightarrow \xi = 0$. Indeed, by using that A is compact and that $|f(\xi)|_p$ is continuous, there exists a constant M' such that $|f(\xi)|_p \geq p^{-M'}$ for $\xi \in A$.

Definition 28 If $f(\xi) \in \mathbb{Z}_p[\xi]$ is an elliptic polynomial of degree d, then we say that $|f|_p^\beta$ is an elliptic symbol, and that $f(\partial, \beta)$ is an elliptic pseudodifferential operator of order d.

By Lemma 24, the Taibleson operator is elliptic for $p \neq 2$. However, there are elliptic symbols which are not radial functions. For instance, $\left|\xi_1^2 - p\xi_2^2\right|_p^\beta = \left[\max\left\{|\xi_1|_p^2, p^{-1}|\xi_2|_p^2\right\}\right]^\beta$. Then, there are two different generalizations of the Taibleson operator (or Vladimirov operator): the W operators which are pseudodifferential operators with radial symbols, and the elliptic operators which include pseudodifferential operators with non-radial symbols.

2.3.2 Decaying of the Fundamental Solution at Infinity

Lemma 29 *For every $t > 0$, $|Z(x,t)| \leq Ct^{\frac{-n}{d\beta}}$, where C is a positive constant. Furthermore, $e^{-t|f(\xi)|_p^\beta} \in L^1$ as a function of ξ, for every $t > 0$.*

Proof Let an integer m be such that $p^{m-1} \leq (Ct)^{\frac{1}{d\beta}} \leq p^m$. By applying (2.23),

$$|Z(x,t)| \leq \int_{\mathbb{Q}_p^n} e^{-Ct\|\xi\|_p^{d\beta}} d^n\xi \leq \int_{\mathbb{Q}_p^n} e^{-p^{(m-1)d\beta}\|\xi\|_p^{d\beta}} d^n\xi$$

$$= \int_{\mathbb{Q}_p^n} e^{-\|p^{-(m-1)}\xi\|_p^{d\beta}} d^n\xi \leq p^n C^{-\frac{n}{d\beta}} \left(\int_{\mathbb{Q}_p^n} e^{-\|z\|_p^{d\beta}} d^n z\right) t^{-\frac{n}{d\beta}} .$$

The result follows from the fact that $e^{-\|z\|_p^{d\beta}}$ is an integrable function. ∎

Define

$$Z_L(x,t,f,\beta) := Z_L(x,t) = \int_{(p^{-L}\mathbb{Z}_p)^n} \chi_p(-x \cdot \xi) \, e^{-t|f(\xi)|_p^\beta} d^n\xi, \ L \in \mathbb{N},$$

where $\beta > 0, t > 0,$ and $f(\xi) \in \mathbb{Z}_p[\xi_1, \ldots, \xi_n]$ is an elliptic polynomial of degree d.

Lemma 30 *If $\|x\|_p \geq p^{M+1}$ and $tp^{Md\beta} \|x\|_p^{-d\beta} \leq 1$, where M is the constant defined in Lemma 26, then there exists a positive constant C such that*

$$|Z_0(x,t)| \leq Ct \|x\|_p^{-d\beta-n}.$$

Proof By applying Fubini's Theorem,

$$Z_0(x,t) = \sum_{l=0}^{\infty} \frac{(-1)^l}{l!} t^l \int_{\mathbb{Z}_p^n} \chi_p(-x \cdot \xi) |f(\xi)|_p^{\beta l} d^n\xi. \qquad (2.25)$$

By using the fact that $\|x\|_p > p^{M+1} > 1$, $\int_{\mathbb{Z}_p^n} \chi_p(-x \cdot \xi) \, d\xi = 0$, and thus (2.25) can be rewritten as

$$Z_0(x,t) = \sum_{l=1}^{\infty} \frac{(-1)^l}{l!} t^l \int_{\mathbb{Z}_p^n} \chi_p(-x \cdot \xi) |f(\xi)|_p^{\beta l} d^n\xi. \qquad (2.26)$$

We set

$$I(j,l) := I(x,f,\beta,j,l) = \int_{\mathbb{Z}_p^n} \chi_p(-p^j x \cdot \xi) |f(\xi)|_p^{\beta l} d^n\xi, \ \text{for } j \geq 0, l \geq 1,$$

and

$$\tilde{I}(j,l,S_0^n) := \tilde{I}(x,f,\beta,j,l,S_0^n) = \int_{S_0^n} \chi_p(-p^j x \cdot \xi) |f(\xi)|_p^{\beta l} d^n\xi,$$

for $j \geq 0, l \geq 1$. By decomposing $\mathbb{Z}_p^n$ as the disjoint union of $(p\mathbb{Z}_p)^n$ and S_0^n,

$$I(0,l) = \int_{\mathbb{Z}_p^n} \chi_p(-x \cdot \xi) |f(\xi)|_p^{\beta l} d\xi$$

$$= \int_{(p\mathbb{Z}_p)^n} \chi_p(-x \cdot \xi) |f(\xi)|_p^{\beta l} d\xi + \int_{S_0^n} \chi_p(-x \cdot \xi) |f(\xi)|_p^{\beta l} d^n\xi$$

$$= p^{-n-\beta dl} I(1,l) + \tilde{I}(0,l,S_0^n).$$

By iterating this formula k-times, we obtain

$$I(0, l) = \sum_{j=0}^{k} p^{-j(n+\beta dl)} \tilde{I}(j, l, S_0^n) + p^{-(k+1)(n+\beta dl)} I(k+1, l).$$

Hence $I(0, l)$ admits the expansion

$$I(0, l) = \sum_{j=0}^{\infty} p^{-j(n+\beta dl)} \tilde{I}(j, l, S_0^n). \tag{2.27}$$

On the other hand, since S_0^n is open compact and f is elliptic, by applying Lemma 26, we obtain

$$\tilde{I}(j, l, A) = \sum_{i=1}^{L_0} p^{-Mn} \chi_p\left(-p^j x \cdot \tilde{\xi}_i\right) \left| f\left(\tilde{\xi}_i\right)\right|_p^{\beta l} \int_{\mathbb{Z}_p^n} \chi_p\left(-p^{j+M} x \cdot y\right) d^n y, \tag{2.28}$$

Now by using

$$\int_{\mathbb{Z}_p^n} \chi_p\left(-p^{j+M} x \cdot y\right) d^n y = \begin{cases} 0 \text{ if } j < -M - ord(x) \\ \\ 1 \text{ if } j \geq -M - ord(x), \end{cases}$$

with $ord(x) = \min_i ord(x_i)$, we can rewrite $\tilde{I}(j, l, A)$ as

$$\begin{cases} p^{-Mn} \sum_{i=1}^{L_0} \chi_p\left(-p^j x \cdot \tilde{\xi}_i\right) \left| f\left(\tilde{\xi}_i\right)\right|_p^{\beta l} & \text{if } j \geq -M - ord(x) \\ \\ 0 & \text{otherwise.} \end{cases} \tag{2.29}$$

We set $\alpha := \alpha(x) = -M - ord(x) \geq 1$ because $\|x\|_p = p^{-ord(x)} \geq p^{M+1}$. With this notation, by combining (2.27)–(2.29) and using that $f(\xi)$ has coefficients in $\mathbb{Z}_p$ and $\tilde{\xi}_i \in \mathbb{Z}_p^n$, $i = 1, \ldots, l$,

$$|I(0, l)| \leq p^{-Mn} \left(\sum_{i=1}^{L_0} \left| f\left(\tilde{\xi}_i\right)\right|_p^{\beta l}\right) \sum_{j=\alpha}^{\infty} p^{-j(n+\beta dl)}$$

$$\leq \left(\frac{L_0}{1 - p^{-(n+\beta d)}}\right) \|x\|_p^{-n} p^{-\alpha \beta l d}.$$

By using this estimation for $|I(0,l)|$ in (2.26),

$$|Z_0(x,t)| \leq \left(\frac{L_0}{1 - p^{-(n+\beta d)}} \right) \|x\|_p^{-n} \left(e^{p^{M d \beta} t \|x\|^{-d\beta}} - 1 \right),$$

finally, by using the hypothesis $t p^{M d \beta} \|x\|_p^{-d\beta} \leq 1$, we have

$$|Z_0(x,t)| \leq Ct \|x\|_p^{-d\beta - n}.$$

∎

Proposition 31 *If* $p^{M d \beta} t \|x\|_p^{-d\beta} \leq 1$*, then* $|Z(x,t)| \leq Ct \|x\|_p^{-d\beta - n}$*, for* $x \in \mathbb{Q}_p^n$ *and* $t > 0$.

Proof By Lemma 29, $\chi_p(-x \cdot \xi) e^{-t|f(\xi)|_p^{\beta}} \in L^1$ as a function of ξ, for $x \in \mathbb{Q}_p^n$ and $t > 0$ fixed. Then, by using the dominated convergence theorem,

$$Z(x,t) = \lim_{L \to \infty} Z_L(x,t) = \lim_{L \to \infty} \int_{(p^{-L}\mathbb{Z}_p)^n} \chi_p(-x \cdot \xi) e^{-t|f(\xi)|_p^{\beta}} d^n \xi.$$

By a change of variables we have

$$Z_L(x,t) = p^{Ln} \int_{\mathbb{Z}_p^n} \chi_p(-p^{-L}x \cdot \xi) e^{-p^{L\beta d}t|f(\xi)|_p^{\beta}} d\xi = p^{Ln} Z_0 \left(p^{-L}x, p^{L\beta d}t \right).$$

Now by applying the Lemma 30,

$$|Z_L(x,t)| \leq Cp^{Ln} \left(\frac{t p^{L\beta d}}{\|x p^{-L}\|_p^{n+\beta d}} \right) \leq C \frac{t}{\|x\|_p^{n+\beta d}},$$

where C is a constant independent of L. Therefore

$$|Z(x,t)| = \lim_{L \to \infty} |Z_L(x,t)| \leq Ct \|x\|_p^{-n-\beta d},$$

if $p^{M d \beta} t \|x\|_p^{-d\beta} \leq 1$. ∎

Theorem 32 *For any* $x \in \mathbb{Q}_p^n$ *and any* $t > 0$,

$$|Z(x,t)| \leq At \left(\|x\|_p + t^{\frac{1}{\beta d}} \right)^{-d\beta - n},$$

where A *is a positive constant.*

Proof If $t^{\frac{1}{\beta d}} \le p^{-M} \|x\|_p$, then $t^{\frac{1}{\beta d}} \le \|x\|_p$ because $M \ge 1$, and by applying Proposition 31,

$$|Z(x,t)| \le Ct \|x\|_p^{-d\beta-n} \le Ct \left(\frac{1}{2} \|x\|_p + \frac{1}{2} t^{\frac{1}{\beta d}} \right)^{-d\beta-n}$$

$$\le \frac{2^{d\beta+n} Ct}{\left(\|x\|_p + t^{\frac{1}{\beta d}} \right)^{d\beta+n}}.$$

Now if $\|x\|_p < t^{\frac{1}{\beta d}}$, by applying Lemma 29,

$$2^{d\beta+n} Ct \left(\|x\|_p + t^{\frac{1}{\beta d}} \right)^{-d\beta-n} \ge Ct^{-\frac{n}{\beta d}} \ge |Z(x,t)|.$$

∎

When considering $Z(x,t)$ as a function of x for t fixed we will write $Z_t(x)$ as before.

Corollary 33 *With the hypothesis of Theorem 32, the following assertions hold: (i) $Z_t(x) \in L^\rho \left(\mathbb{Q}_p^n \right)$, for $1 \le \rho \le \infty$, for $t > 0$; (ii) $Z_t(x)$ is a continuous function in x, for $t > 0$ fixed.*

Proof (i) The first part follows directly from the estimation given in Theorem 32. (ii) The continuity follows from the fact that $Z_t(x)$ is the Fourier transform of $e^{-t|f(\xi)|_p^\beta}$, $t > 0$, that is an integrable function by Lemma 25. ∎

2.3.3 Positivity of the Fundamental Solution

Theorem 34 $Z(x,t) \ge 0$ *for every* $x \in \mathbb{Q}_p^n$ *and every* $t > 0$.

Proof We start by making the following observation about the fiber of $f : \mathbb{Q}_p^n \to \mathbb{Q}_p$ at $\lambda \in \mathbb{Q}_p$.

(**Claim A**) $f^{-1}(\lambda)$ is a compact subset of $\mathbb{Q}_p^n$.
Since f is continuous $f^{-1}(\lambda)$ is a closed subset of $\mathbb{Q}_p^n$. By applying (2.23),

$$f^{-1}(\lambda) \subseteq \left\{ \xi \in \mathbb{Q}_p^n; \|\xi\|_p \le \left(\frac{|\lambda|}{C_0} \right)^{\frac{1}{d}} \right\},$$

and thus $f^{-1}(\lambda)$ is a bounded subset of $\mathbb{Q}_p^n$.
(**Claim B**) The critical set $C_f = \{\xi \in \mathbb{Q}_p^n; \nabla f(\xi) = 0\}$ of the mapping f is reduced to the origin of $\mathbb{Q}_p^n$.

This claim follows from the Euler identity

$$\frac{1}{d} \sum_{i=1}^{n} \xi_i \frac{\partial f(\xi)}{\partial \xi_i} = f(\xi),$$

and the fact that f is an elliptic polynomial.

On the other hand, since $\chi_p(-x \cdot \xi) 1_{f^{-1}(\lambda)}(\xi) \in \mathcal{D}(\mathbb{Q}_p^n)$, as a function of ξ, for $\lambda \neq 0$, by applying integration on fibers to $Z(x,t)$, see Chap. 1, formula (1.3), with $t > 0$ fixed,

$$Z(x,t) = \int_{\mathbb{Q}_p \setminus \{0\}} e^{-t|\lambda|_p^{\beta}} \left(\int_{f(\xi)=\lambda} \chi_p(-x \cdot \xi) |\gamma_{\mathrm{GL}}| \right) d\lambda,$$

where $|\gamma_{\mathrm{GL}}|$ is the measure induced by the Gel'fand-Leray form along the fiber $f^{-1}(\lambda)$. Hence in order to prove the theorem, it is sufficient to show that

$$F(\lambda, x) := \left(\int_{f(\xi)=\lambda} \chi_p(-x \cdot \xi) |\gamma_{\mathrm{GL}}| \right) \geq 0, \text{ for every } x \in \mathbb{Q}_p^n \setminus \{0\}.$$

Let $\tilde{\xi}$ be a fixed point of $f^{-1}(\lambda)$, $\lambda \in \mathbb{Q}_p \setminus \{0\}$. By Claim B we may assume, after renaming the variables if necessary, that $\frac{\partial f}{\partial \xi_n}(\tilde{\xi}) \neq 0$. We set $y = \phi(\xi)$ with

$$y_j := \begin{cases} \xi_j & j = 1, \ldots, n-1 \\ f(\tilde{\xi} + p^e \xi) - f(\tilde{\xi}) & j = n. \end{cases}$$

By applying the non-Archimedean implicit function theorem (see Chap. 1, Theorem 1) there exist $e, l \in \mathbb{N}$ such that $y = \phi(\xi)$ is a bianalytic mapping from $\mathbb{Z}_p^n$ onto $\left(p^l \mathbb{Z}_p\right)^n$. Then

$$\xi = \phi^{-1}(y) = \left(y_1, \ldots, y_{n-1}, \sum_{j=1}^{\infty} G_j(y) \right),$$

where $G_j(y)$ is a form of degree j, and $G_1(y) \neq 0$. By shrinking the neighborhoods around $\tilde{\xi}$ and the origin, i.e., by taking e and l big enough, we may assume that the following conditions hold:

(C) the Jacobian $J_{\phi^{-1}}$ of ϕ^{-1} satisfies $\left| J_{\phi^{-1}}(y) \right|_p = \left| J_{\phi^{-1}}(0) \right|_p$, for every $y \in \left(p^l \mathbb{Z}_p \right)^n$;

(D) $ord\left(x_n \sum_{j=1}^{\infty} G_j(y)\right) \geq 0$, for any $y \in \left(p'\mathbb{Z}_p\right)^n$.

Since $f^{-1}(\lambda)$, $\lambda \in \mathbb{Q}_p \setminus \{0\}$, is a compact subset by Claim A, $F(\lambda, x)$ can expressed as a finite sum of integrals of the form

$$\int_{\tilde{\xi}+p^e\mathbb{Z}_p^n \cap f^{-1}(\lambda)} \chi_p(-x \cdot \xi) |\gamma_{GL}|.$$

Now by changing variables $\xi = \phi^{-1}(y)$, and using (C), (D), we obtain

$$\int_{\tilde{\xi}+p^e\mathbb{Z}_p^n \cap f^{-1}(\lambda)} \chi_p(-x \cdot \xi) |\gamma_{GL}|$$

$$= |J_{\phi^{-1}}(0)|_p \int_{p'\mathbb{Z}_p^{n-1}} \chi_p\left(-\sum_{j=1}^{n-1} x_j\xi_j - x_n \sum_{l=1}^{\infty} G_j(y)\right) d^{n-1}y$$

$$= |J_{\phi^{-1}}(0)|_p \int_{p'\mathbb{Z}_p^{n-1}} \chi_p\left(-\sum_{j=1}^{n-1} x_j\xi_j\right) d^{n-1}y$$

$$= \left(p^{-l(n-1)} |J_{\phi^{-1}}(0)|_p\right) 1_{p^{-l}\mathbb{Z}_p^{n-1}}(x) \geq 0,$$

where $1_{p^{-l}\mathbb{Z}_p^{n-1}}(x)$ denotes the characteristic function of $p^{-l}\mathbb{Z}_p^{n-1}$. Therefore $F(\lambda, x) \geq 0$. ∎

2.3.4 Some Additional Results

We denote by $C_b := C_b\left(\mathbb{Q}_p^n, \mathbb{R}\right)$ the $\mathbb{R}$-vector space of all functions $\varphi : \mathbb{Q}_p^n \to \mathbb{R}$ which are continuous and satisfy $\|\varphi\|_{L^\infty} = \sup_{x \in \mathbb{Q}_p^n} |\varphi(x)| < \infty$.

Proposition 35 *The fundamental solution has the following properties:*

(i) $\int_{\mathbb{Q}_p^n} Z(x, t) \, d^n x = 1$, *for any* $t > 0$;

(ii) *if* $\varphi \in C_b$, *then* $\lim_{(x,t)\to(x_0,0)} \int_{\mathbb{Q}_p^n} Z(x - y, t) \varphi(y) \, d^n y = \varphi(x_0)$;

(iii) $Z(x, t + t') = \int_{\mathbb{Q}_p^n} Z(x - y, t) Z(y, t') \, d^n y$, *for* $t, t' > 0$.

Proof

(i) It follows from Corollary 33 and the Fourier inversion formula.
(ii) We set $u(x,t) = \int_{\mathbb{Q}_p^n} Z(x-y,t)\,\varphi(y)\,d^n y$. We have to show that

$$\lim_{(x,t)\to(x_0,0)} u(x,t) = \varphi(x_0),$$

for any fixed $x_0 \in \mathbb{Q}_p^n$. Since φ is continuous at x_0 there exists a ball $B_{-e}^n(x_0) = \{y \in \mathbb{Q}_p^n; \|y - x_0\|_p \le p^{-e}\}$, such that $|\varphi(y) - \varphi(x_0)| < \frac{\epsilon}{2}$, for every $y \in B_{-e}^n(x_0)$. Then $|u(x,t) - \varphi(x_0)| \le |I_1| + |I_2|$, where

$$|I_1| := \left| \int\limits_{\|y-x_0\|_p \le p^{-e}} Z(x-y,t)\,[\varphi(y) - \varphi(x_0)]\,d^n y \right|,$$

$$|I_2| := \left| \int\limits_{\|y-x_0\|_p > p^{-e}} Z(x-y,t)\,[\varphi(y) - \varphi(x_0)]\,d^n y \right|.$$

By using the continuity of φ and (i),

$$|I_1| < \frac{\epsilon}{2}, \text{ for } y \in B_{-e}^n(x_0).$$

By applying Theorem 32 to $|I_2|$,

$$|I_2| \le 2Ct\,\|\varphi\|_{L^\infty} \int\limits_{\|y-x_0\|_p > p^{-e}} \|x-y\|_p^{-d\beta-n}\,d^n y.$$

Now, since we are interested in the values of x close to x_0, we may assume that $\|x - x_0\|_p < p^{-e}$, then by the ultrametric triangle inequality,

$$\|x - y\|_p = \max\left(\|x - x_0\|_p, \|y - x_0\|_p\right) = \|y - x_0\|_p,$$

and

$$|I_2| \le 2Ct\,\|\varphi\|_{L^\infty} \int\limits_{\|z\|_p > p^{-e}} \|z\|_p^{-d\beta-n}\,d^n z \le C_1 t\,\|\varphi\|_{L^\infty},$$

for $t > 0$, where C_1 is a positive constant. Note that $\|\varphi\|_{L^\infty} = 0$, implies $\varphi \equiv 0$, since φ is a continuous function. In this case the theorem is valid. For this reason we assume that $\|\varphi\|_{L^\infty} > 0$. Hence

$$|I_2| < \frac{\epsilon}{2}, \text{ for } (t,x) \text{ satisfying } \|x - x_0\|_p < p^{-e}, 0 < t < \frac{\varepsilon}{2C_1\,\|\varphi\|_{L^\infty}}.$$

(iii) By using that $e^{-t|f(\xi)|_p^\beta} \in L^1$, for every $t > 0$,

$$\int_{\mathbb{Q}_p^n} Z(x - y, t) Z(y, t') \, d^n y = \mathcal{F}^{-1}\left(e^{-t|f(\xi)|_p^\beta} e^{-t'|f(\xi)|_p^\beta}\right) = Z(x, t + t'),$$

for $t, t' > 0$.

∎

2.3.5 The Cauchy Problem

In this section we study the following Cauchy problem:

$$\begin{cases} \frac{\partial u(x,t)}{\partial t} + (f(\partial, \beta) u)(x, t) = 0, \ t > 0 \\[2mm] u(x, 0) = \varphi(x), \end{cases} \tag{2.30}$$

where $\varphi \in L^1 \cap C_b$.

Lemma 36 *If $\varphi \in L^1$, then the function*

$$u(x, t) = \int_{\mathbb{Q}_p^n} Z(x - y, t) \varphi(y) \, d^n y \tag{2.31}$$

is a classical solution of the equation

$$\frac{\partial u(x, t)}{\partial t} + (f(\partial, \beta) u)(x, t) = 0, t > 0.$$

In addition, $u(x, t) \in L^\rho$, for $1 \le \rho \le \infty$, for every fixed $t > 0$.

Proof It is clear that one may differentiate in (2.31) under the integral sign:

$$\frac{\partial u(x, t)}{\partial t} = \int_{\mathbb{Q}_p^n} \varphi(y) \frac{\partial}{\partial t} Z(x - y, t) \, d^n y = \frac{\partial Z(x, t)}{\partial t} * \varphi(x), \text{ for } t > 0. \tag{2.32}$$

On the other hand, since $Z(x, t) \in L^\rho$, $1 \le \rho \le \infty$, for any fixed $t > 0$ (cf. Corollary 33), and $\varphi \in L^1$, then $u(x, t) \in L^\rho$, $1 \le \rho \le \infty$, for any fixed $t > 0$, and its Fourier transform with respect x is $e^{-t|f(\xi)|_p^\beta} (\mathcal{F}\varphi)(\xi) \in L^\rho$, $1 \le \rho \le \infty$, because $(\mathcal{F}\varphi)(\xi) \in L^\infty$, by the Riemann-Lebesgue Theorem, and the fact that f is elliptic. Now by using Lemma 25 we have $|f(\xi)|_p^\beta e^{-t|f(\xi)|_p^\beta} \in L^1 \cap L^2$ for any fixed $t > 0$.

Then $(f(\partial, \beta) u_0)(x, t)$ is given by

$$(f(\partial, \beta) u)(x, t) = \mathcal{F}_{\xi \to x}^{-1} \left(|f(\xi)|_p^{\beta} \, e^{-t|f(\xi)|_p^{\beta}} \right) * \varphi(x)$$

$$= -\mathcal{F}_{\xi \to x}^{-1} \left(\frac{\partial}{\partial t} e^{-t|f(\xi)|_p^{\beta}} \right) * \varphi(x),$$

for $t > 0$, and since one may differentiate in (2.21) under the integral,

$$(f(\partial, \beta) u)(x, t) = -\frac{\partial Z(x, t)}{\partial t} * \varphi(x). \tag{2.33}$$

Now the result follows directly from (2.32) and (2.33). ∎

Lemma 37 *Let $u(x, t)$ be as in Lemma 36. Then the following assertions hold: (i) $u(x, t)$ is continuous for any $t \geq 0$;(ii) $|u(x, t)| \leq \|\varphi\|_{L^\infty}$ for any $t \geq 0$.*

Proof (i) For $t > 0$, since $|f(\xi)|_p^{\beta} \, e^{-t|f(\xi)|_p^{\beta}} (\mathcal{F}\varphi)(\xi) \in L^1$, $u(x, t) = \mathcal{F}_{\xi \to x}^{-1} \left(|f(\xi)|_p^{\beta} \, e^{-t|f(\xi)|_p^{\beta}} (\mathcal{F}\varphi)(\xi) \right)$ is continuous. The continuity at $t = 0$ follows from the fact that $u(x, 0) = \varphi(x) = \lim_{t \to 0} u(x, 0)$, cf. Proposition 35 (ii). For $t > 0$, the result follows from the Young inequality. ∎

Theorem 38 *If $\varphi \in L^1 \cap C_b$, then the Cauchy problem*

$$\begin{cases} \dfrac{\partial u(x,t)}{\partial t} + (f(\partial, \beta) u)(x, t) = 0, \; x \in \mathbb{Q}_p^n, \; t > 0 \\[2mm] u(x, 0) = \varphi(x) \end{cases}$$

has a classical solution given by

$$u(x, t) = \int_{\mathbb{Q}_p^n} Z(x - y, t) \, \varphi(y) \, d^n y.$$

Furthermore, the solution has the following properties:

(1) $u(x, t)$ is a continuous function in x, for every fixed $t \geq 0$;
(2) $\sup_{(x,t) \in \mathbb{Q}_p^n \times [0, +\infty)} |u(x, t)| \leq \|\varphi\|_{L^\infty}$;
(3) $u(x, t) \in L^\rho$, $1 \leq \rho \leq \infty$, for any fixed $t > 0$.

Proof The result follows from Lemmas 36, 37. ∎

2.3.6 Markov Processes Over $\mathbb{Q}_p^n$

Theorem 39 *$Z(x, t)$ is the transition density of a time and space homogeneous Markov process which is bounded, right-continuous and has no discontinuities other than jumps.*

Proof By Proposition 35 (iii) the family of operators

$$(\Theta\,(t)f)\,(x) = \int\limits_{\mathbb{Q}_p^n} Z\,(x-y,t)\,f\,(y)\,d^n y$$

has the semigroup property. We know that $Z\,(x,t)\,\geq\,0$ and $\Theta\,(t)$ preserves the function $f\,(x)\,\equiv\,1$ (cf. Proposition 35 (i)). Thus $\Theta\,(t)$ is a Markov semigroup. The requiring properties of the corresponding Markov process follow from Theorem 32 and general theorems of the theory of Markov processes [39], see also Sect. 2.2.5.∎

Remark 40 By using the results of [42], it is possible to show that there exists a Lévy process with state space $\mathbb{Q}_p^n$ and transition function

$$P(t,x,E) = \begin{cases} Z_t(x) * 1_E(x) \text{ for } t > 0, x \in \mathbb{Q}_p^n \\ \\ 1_E(x) \qquad\qquad \text{for } t = 0, x \in \mathbb{Q}_p^n, \end{cases}$$

where E is an element of the family of subsets of $\mathbb{Q}_p^n$ formed by finite unions of disjoint balls and the empty set. However, for the sake of simplicity we state our results in the framework of Markov processes.

Chapter 3
Non-Archimedean Parabolic-Type Equations with Variable Coefficients

3.1 Introduction

This chapter is devoted to the study of a class of parabolic-type equations with variable coefficients, which generalizes the parabolic-type equations attached to the W operators. The theory of these equations was initiated by Kochubei in [81], see also [80, Chapter 4]. He studied, in dimension one, p-adic parabolic-type equations with variable coefficients and their associated Markov processes. Later Rodríguez-Vega in [97] extended some of the results of Kochubei to the n-dimensional case by using the Taibleson operator instead of the Vladimirov operator. Building up on [25] and [80], Chacón-Cortes and Zúñiga-Galindo developed a theory of p-adic parabolic-type equations with variable coefficients attached to operators W, which contains the one-dimensional p-adic heat equation of [111], the equations studied by Kochubei in [80], and the equations studied by Rodríguez-Vega in [97]. In this chapter, we establish the existence and uniqueness of solutions for the Cauchy problem for these equations. We show that the fundamental solutions of these equations are transition density functions of Markov processes, and finally, we study the well-possedness of the Cauchy problem.

3.2 A Class of Non-local Operators

We recall that a complex-valued function φ defined on $\mathbb{Q}_p^n$ is *called locally constant* if for any $x \in \mathbb{Q}_p^n$ there exists an integer $l = l(x) \in \mathbb{Z}$ such that

$$\varphi(x + x') = \varphi(x) \text{ for } x' \in B_l^n. \tag{3.1}$$

© Springer International Publishing AG 2016
W.A. Zúñiga-Galindo, *Pseudodifferential Equations Over Non-Archimedean Spaces*, Lecture Notes in Mathematics 2174,
DOI 10.1007/978-3-319-46738-2_3

The set of all locally constant functions φ, for which the integer $l(x)$ is independent of x, form $\mathbb{C}$-vector space denoted by $\tilde{\mathcal{E}}(\mathbb{Q}_p^n) := \tilde{\mathcal{E}}$. Given $\varphi \in \tilde{\mathcal{E}}$, we call the largest possible $l = l(\varphi)$, the *parameter of local constancy of* φ.

Denote by $\mathfrak{M}_\lambda$, with $\lambda \geq 0$, the $\mathbb{C}$-vector space of all the functions $\varphi \in \tilde{\mathcal{E}}$ satisfying $|\varphi(x)| \leq C(1 + \|x\|_p^\lambda)$. If the function φ depends also on a parameter t, we shall say that φ belongs to $\mathfrak{M}_\lambda$ *uniformly with respect to* t, if its constant C and its parameter of local constancy do not depend on t. Notice that, if $0 \leq \lambda_1 \leq \lambda_2$, then $\mathfrak{M}_0 \subseteq \mathfrak{M}_{\lambda_1} \subseteq \mathfrak{M}_{\lambda_2}$, and that $\mathcal{D}(\mathbb{Q}_p^n) \subseteq \mathfrak{M}_0$.

Take $\mathbb{R}_+ = \{x \in \mathbb{R}; x \geq 0\}$ as before, and fix a function $w_\alpha : \mathbb{Q}_p^n \to \mathbb{R}_+$ having the following properties:

(i) $w_\alpha(y)$ is a radial (i.e. $w_\alpha(y) = w_\alpha(\|y\|_p)$), continuous and increasing function;
(ii) $w_\alpha(y) = 0$ if and only if $y = 0$;
(iii) there exist constants $C_0, C_1 > 0$, and $\alpha > n$ such that

$$C_0 \|y\|_p^\alpha \leq w_\alpha(\|y\|_p) \leq C_1 \|y\|_p^\alpha \text{ for any } y \in \mathbb{Q}_p^n.$$

Set

$$A_{w_\alpha}(\xi) := \int_{\mathbb{Q}_p^n} \frac{1 - \chi_p(-y \cdot \xi)}{w_\alpha(\|y\|_p)} d^n y.$$

In Sect. 2.2 of Chap. 2, we establish that function A_{w_α} is radial, positive, continuous, $A_{w_\alpha}(0) = 0$, and $A_{w_\alpha}(\xi) = A_{w_\alpha}(\|\xi\|_p) = A_{w_\alpha}(p^{-ord(\xi)})$ is a decreasing function of $ord(\xi)$. In addition, we introduce the following operator:

$$(W_\alpha \varphi)(x) = \int_{\mathbb{Q}_p^n} \frac{\varphi(x - y) - \varphi(x)}{w_\alpha(\|y\|_p)} d^n y, \; \varphi \in \mathcal{D}(\mathbb{Q}_p^n). \tag{3.2}$$

Lemma 41 *If $\alpha - n > \lambda$, then W_α can be extended to $\mathfrak{M}_\lambda$ and formula (3.2) holds. Furthermore, $W_\alpha : \mathfrak{M}_\lambda \to \mathfrak{M}_\lambda$.*

Proof Notice that if $\varphi \in \mathfrak{M}_\lambda$, there exists a constant $l = l(\varphi) \in \mathbb{Z}$, such that

$$(W_\alpha \varphi)(x) = \int_{\|y\|_p \geq p^l} \frac{\varphi(x - y) - \varphi(x)}{w_\alpha(\|y\|_p)} d^n y. \tag{3.3}$$

We now show that $|(W_\alpha \varphi)(x)| \leq A(1 + \|x\|_p^\lambda)$. By using that $\varphi \in \mathfrak{M}_\lambda$, and $\alpha > n$,

$$|(W_\alpha \varphi)(x)| \leq C \int_{\|y\|_p \geq p^l} \frac{(1 + \|x - y\|_p^\lambda)}{\|y\|_p^\alpha} d^n y + C'(1 + \|x\|_p^\lambda).$$

Hence, it is sufficient to show that the above integral can be bounded by $A(1+\|x\|_p^\lambda)$, for some positive constant A . If $\|x\|_p > \|y\|_p$,

$$\int\limits_{\|y\|_p \geq p^l} \frac{(1+\|x-y\|_p^\lambda)}{\|y\|_p^\alpha} d^n y \leq (1+\|x\|_p^\lambda) \int\limits_{\|y\|_p \geq p^l} \frac{1}{\|y\|_p^\alpha} d^n y$$

$$= B(1 + \|x\|_p^\lambda),$$

where B is a positive constant. If $\|x\|_p < \|y\|_p$, by using $\alpha - n > \lambda$,

$$\int\limits_{\|y\|_p \geq p^l} \frac{(1+\|x-y\|_p^\lambda)}{\|y\|_p^\alpha} d^n y \leq \int\limits_{\|y\|_p \geq p^l} \frac{(1+\|y\|_p^\lambda)}{\|y\|_p^\alpha} d^n y < \infty.$$

If $\|x\|_p = \|y\|_p \geq p^l$, we take $x = p^L u$, $y = p^L v$, with $\|v\|_p = \|u\|_p = 1, L \in \mathbb{Z}$, then

$$\int\limits_{\|y\|_p = \|x\|_p} \frac{(1+\|x-y\|_p^\lambda)}{\|y\|_p^\alpha} d^n y = p^{-L(n-\alpha)} \int\limits_{\|v\|_p = 1} (1 + p^{-L\lambda} \|u - v\|_p^\lambda) d^n v$$

$$\leq A \left(\|x\|_p^{-(\alpha-n)} + \|x\|_p^{-(\alpha-n-\lambda)} \right) \leq A'(p, l, \alpha, n, \lambda),$$

where A, A' are positive constants. Finally, by (3.3) $W_\alpha \varphi$ is locally constant. ∎

3.3 Parabolic-Type Equations with Constant Coefficients

Consider the following Cauchy problem:

$$\begin{cases} \frac{\partial u}{\partial t}(x, t) - \kappa(W_\alpha u)(x, t) = f(x, t), \ x \in \mathbb{Q}_p^n, t \in (0, T] \\ u(x, 0) = \varphi(x), \end{cases} \tag{3.4}$$

where, $\alpha > n$, κ, T are positive constants, $\varphi \in Dom(W_\alpha) := \mathfrak{M}_\lambda$, with $\alpha - n > \lambda$, f is continuous in (x, t) and belongs to $\mathfrak{M}_\lambda$ uniformly with respect to t, and $u : \mathbb{Q}_p^n \times [0, T] \to \mathbb{C}$ is an unknown function. We say that $u(x, t)$ is a *solution of* (3.4), if $u(x, t)$ is continuous in (x, t), $u(\cdot, t) \in Dom(W_\alpha)$ for $t \in [0, T]$, $u(x, \cdot)$ is continuously differentiable for $t \in (0, T]$, $u(x, t) \in \mathfrak{M}_\lambda$ uniformly in t, and u satisfies (3.4) for all $t > 0$.

Cauchy problem (3.4) was studied in Chap. 2 using semigroup theory, this approach cannot be used in the space $\mathfrak{M}_\lambda$, since it is not contained in L^ρ for any $\rho \in [1, \infty]$.

We define

$$Z(x, t; w_\alpha, \kappa) := Z(x, t) = \int_{\mathbb{Q}_p^n} e^{-\kappa t A_{w_\alpha}(\|\xi\|_p)} \chi_p(x \cdot \xi) d^n \xi, \qquad (3.5)$$

for $t > 0$ and $x \in \mathbb{Q}_p^n$. Notice that, $Z(x, t) = \mathcal{F}_{\xi \to x}^{-1}[e^{-\kappa t A_{w_\alpha}(\|\xi\|_p)}] \in L^1 \cap L^2$ for $t > 0$, since $C' \|\xi\|_p^{\alpha - n} \leq A_{w_\alpha}(\|\xi\|_p) \leq C'' \|\xi\|_p^{\alpha - n}$, cf. Chap. 2, Lemma 8. Furthermore, $Z(x, t) \geq 0$, for $t > 0$, $x \in \mathbb{Q}_p^n$, cf. Chap. 2, Lemma 12. These functions are called *heat kernels*. When considering $Z(x, t)$ as a function of x for t fixed we will write $Z_t(x)$.

We set

$$u_1(x, t) := \int_{\mathbb{Q}_p^n} Z(x - y, t) \varphi(y) d^n y$$

and

$$u_2(x, t) := \int_0^t \int_{\mathbb{Q}_p^n} Z(x - y, t - \theta) f(y, \theta) d^n y d\theta,$$

for $\varphi, f \in \mathfrak{M}_\lambda$ with $\alpha - n > \lambda$, for $0 \leq t \leq T$, and $x \in \mathbb{Q}_p^n$. The main result of this section is the following:

Theorem 42 *The function* $u(x, t) = u_1(x, t) + u_2(x, t)$ *is a solution of Cauchy Problem (3.4).*

The proof requires several steps.

3.3.1 Claim $u(x, t) \in \mathfrak{M}_\lambda$

In order to prove this claim, we need some preliminary results.

Remark 43 The function $Z_t(x)$ is radial since it is the inverse Fourier transform of the radial function $e^{-\kappa t A_{w_\alpha}(\|\xi\|_p)}$. Then $Z_t(x)$ is locally constant in $\mathbb{Q}_p^n \setminus \{0\}$. Furthermore, $Z_t(x + y) = Z_t(x)$ if $\|y\|_p < \|x\|_p$ for any $y \in \mathbb{Q}_p^n$ and $x \in \mathbb{Q}_p^n \setminus \{0\}$, and $t > 0$.

Lemma 44 *There exist positive constants* C_1, C_2 *such that* $Z(x, t)$ *satisfies the following conditions:*

(i) $Z(x, t) \leq C_1 t^{-\frac{n}{\alpha - n}}$, *for* $t > 0$ *and* $x \in \mathbb{Q}_p^n$;
(ii) $Z(x, t) \leq C_2 t \|x\|_p^{-\alpha}$, *for* $t > 0$ *and* $x \in \mathbb{Q}_p^n \setminus \{0\}$;

(iii) $Z(x,t) \leq \max\{2^{\alpha}C_1, 2^{\alpha}C_2\}t\left(\|x\|_p + t^{\frac{1}{\alpha-n}}\right)^{-\alpha}$, *for $t > 0$ and $x \in \mathbb{Q}_p^n$;*

(iv) $\int_{\mathbb{Q}_p^n} Z(x,t)d^n x = 1$, *for $t > 0$.*

Proof (i) By (3.5) and Lemma 8 in Chap. 2,

$$Z(x,t) \leq \int\limits_{\mathbb{Q}_p^n} e^{-\kappa t A_{w\alpha}(\|\xi\|_p)} d^n \xi \leq \int\limits_{\mathbb{Q}_p^n} e^{-C_0 t \|\xi\|_p^{\alpha-n}} d^n \xi.$$

Let m be an integer such that $p^{m-1} \leq t^{\frac{1}{\alpha-n}} \leq p^m$, then

$$Z(x,t) \leq \int\limits_{\mathbb{Q}_p^n} e^{-C_0\|p^{-(m-1)}\xi\|_p^{\alpha-n}} d^n \xi,$$

now, by changing variables as $z = p^{-(m-1)}\xi$, we have

$$Z(x,t) \leq p^{-(m-1)n} \int\limits_{\mathbb{Q}_p^n} e^{-C_0\|z\|_p^{\alpha-n}} d^n z \leq C_1 t^{-\frac{n}{\alpha-n}}.$$

(ii) It follows from Lemma 11 in Chap. 2. (iii) The result is obtained from the two following inequalities. If $\|x\|_p \geq t^{\frac{1}{\alpha-n}}$, then $\|x\|_p \geq \frac{\|x\|_p}{2} + \frac{t^{\frac{1}{\alpha-n}}}{2}$ and $\|x\|_p^{-\alpha} \leq 2^{\alpha}\left(\|x\|_p + t^{\frac{1}{\alpha-n}}\right)^{-\alpha}$, multiplying by $C_2 t$ and using (ii),

$$Z(x,t) \leq 2^{\alpha} C_2 t \left(\|x\|_p + t^{\frac{1}{\alpha-n}}\right)^{-\alpha}.$$

If $\|x\|_p \leq t^{\frac{1}{\alpha-n}}$, then $\frac{\|x\|_p}{2} + \frac{t^{\frac{1}{\alpha-n}}}{2} \leq t^{\frac{1}{\alpha-n}}$ and $\left(\|x\|_p + t^{\frac{1}{\alpha-n}}\right)^{-\alpha} \geq 2^{-\alpha}t^{\frac{-\alpha}{\alpha-n}} = 2^{-\alpha}t^{-1-\frac{n}{\alpha-n}}$, multiplying by C_1 and using (i),

$$Z(x,t) \leq 2^{\alpha} C_1 t \left(\|x\|_p + t^{\frac{1}{\alpha-n}}\right)^{-\alpha}.$$

(iv) By (iii), $Z_t(x) \in L^1(\mathbb{Q}_p^n)$ for $t > 0$. Now, the announced identity follows by applying the Fourier inversion formula. $\blacksquare$

Proposition 45 *If $b > 0$, $0 \leq \lambda < \alpha$, and $x \in \mathbb{Q}_p^n$, then*

$$\int\limits_{\mathbb{Q}_p^n} \left(b + \|x - \xi\|_p\right)^{-\alpha-n} \|\xi\|_p^{\lambda} d^n \xi \leq Cb^{-\alpha}\left(1 + \|x\|_p^{\lambda}\right),$$

where the constant C does not depend on b or x.

Proof Let m be an integer such that $p^{m-1} \le b \le p^m$. Then

$$\left(b + ||x - \xi||_p\right)^{-\alpha-n} \le \left(p^{m-1} + ||x - \xi||_p\right)^{-\alpha-n},$$

and

$$I(b, x) \le I(p^{m-1}, x) = \int_{\mathbb{Q}_p^n} \left(p^{m-1}|_p + ||x - \xi||_p\right)^{-\alpha-n} ||\xi||_p^\lambda \, d^n\xi$$

$$= p^{(m-1)(-\alpha-n)} \int_{\mathbb{Q}_p^n} \left(1 + ||p^{m-1}x - p^{m-1}\xi||_p\right)^{-\alpha-n} ||\xi||_p^\lambda \, d^n\xi$$

$$= p^{(m-1)(\lambda-\alpha)} \int_{\mathbb{Q}_p^n} \left(1 + ||p^{m-1}x - \eta||_p\right)^{-\alpha-n} ||\eta||_p^\lambda \, d^n\eta$$

$$= p^{(m-1)(\lambda-\alpha)} I(1, p^{m-1}x). \qquad (3.6)$$

Let $p^{m-1}x = y$, $||y||_p = p^l$. We have

$$I(1, y) = I_1(y) + I_2(y) + I_3(y),$$

where

$$I_1(y) = \sum_{k=-\infty}^{l-1} \int_{||\eta||_p = p^k} \left(1 + ||y - \eta||_p\right)^{-\alpha-n} ||\eta||_p^\lambda \, d^n\eta,$$

$$I_2(y) = \int_{||\eta||_p = p^l} \left(1 + ||y - \eta||_p\right)^{-\alpha-n} ||\eta||_p^\lambda \, d^n\eta,$$

$$I_3(y) = \sum_{k=l+1}^{\infty} \int_{||\eta||_p = p^k} \left(1 + ||y - \eta||_p\right)^{-\alpha-n} ||\eta||_p^\lambda \, d^n\eta.$$

The result follows from the following estimations:

Claim A $I_1(y) \le C_0(1 + ||y||_p)^{-\alpha-n}||y||_p^{\lambda+n}$;

Claim B $I_2(y) \le C_1||y||_p^\lambda$;

Claim C $I_3(y) \le C_2$.

Indeed, from the claims we have $I(1, y) \le C_3(1 + ||y||_p^\lambda)$, and by (3.6),

$$I(b, x) \le C_3 p^{(m-1)(\lambda-\alpha)}(1 + p^{(1-m)\lambda}||x||_p^\lambda)$$

$$\le C_3 p^{-m\alpha}(1 + ||x||_p^\lambda) \le Cb^{-\alpha}\left(1 + ||x||_p^\lambda\right).$$

We now prove the announced claims.

Proof of Claim A

$$I_1(y) = (1 - p^{-n})(1 + ||y||_p)^{-\alpha-n} \sum_{k=-\infty}^{l-1} p^{(\lambda+n)k}$$

$$\leq C_0(1 + ||y||_p)^{-\alpha-n} ||y||_p^{\lambda+n}.$$

Proof of Claim B

$$I_2(y) = ||y||_p^{\lambda} \int_{||p^l \eta||_p = 1} \left(1 + ||y - \eta||_p\right)^{-\alpha-n} ||\eta||_p^{\lambda} \, d^n \eta$$

$$I_2(y) = ||y||_p^{\lambda+n} \int_{||\eta||_p = p^l} \left(1 + ||y - p^{-l}\eta||_p\right)^{-\alpha-n} d^n \eta$$

$$= ||y||_p^{\lambda-\alpha} \int_{||\eta||_p = 1} \left(p^{-l} + ||u - \eta||_p\right)^{-\alpha-n} d^n \eta, \text{ with } u = p^l y.$$

We set

$$A_m = \{\eta \in \mathbb{Q}_p^n; ||\eta||_p = 1 \text{ and } ||u - \eta||_p = p^{-m}\}, \text{ for } m \in \mathbb{N},$$

and for I non-empty subset of $\{1, 2, \ldots, n\}$,

$$A_{m,I} = \{\eta \in A_m; |u_i - \eta_i|_p = p^{-m} \text{ for } i \in I \text{ and } |u_i - \eta_i|_p < p^{-m} \text{ for } i \notin I\},$$

where $u = (u_1, \ldots, u_n)$, $\eta = (\eta_1, \ldots, \eta_n) \in \mathbb{Q}_p^n$, with $||\eta||_p = ||u||_p = 1$. With this notation we have $A_m \subseteq \bigcup_I A_{m,I}$,

$$vol(A_{m,I}) \leq (p^{-m}(1 - p^{-1}))^{|I|}(p^{-m-1})^{n-|I|},$$

here $|I|$ denotes the cardinality of I, then

$$vol(A_m) \leq \sum_{|I|=0}^{n} \binom{n}{|I|} (p^{-m}(1 - p^{-1}))^{|I|}(p^{-m-1})^{n-|I|} = p^{-mn},$$

and

$$I_2(y) = ||y||_p^{\lambda-\alpha} \sum_{m=0}^{\infty} \int_{A_m} \left(p^{-l} + ||u - \eta||_p\right)^{-\alpha-n} d^n \eta$$

$$\leq ||y||_p^{\lambda-\alpha} \sum_{m=0}^{\infty} \left(p^{-l} + p^{-m}\right)^{-\alpha-n} p^{-mn}$$

$$
= \frac{||y||_p^{\lambda-\alpha}}{1-p^{-n}} \sum_{m=0}^{\infty} \int_{||\eta||_p=p^{-m}} \left(p^{-l} + ||\eta||_p\right)^{-\alpha-n} d^n\eta
$$

$$
= \frac{||y||_p^{\lambda-\alpha}}{1-p^{-n}} \int_{||\eta||_p \leq 1} \left(p^{-l} + ||\eta||_p\right)^{-\alpha-n} d^n\eta
$$

$$
\leq C_1' ||y||_p^{\lambda-\alpha} \int_{\mathbb{Q}_p^n} \left(p^{-l} + ||\eta||_p\right)^{-\alpha-n} d^n\eta
$$

$$
= C_1' ||y||_p^{\lambda+n} \int_{\mathbb{Q}_p^n} \left(1 + ||p^l\eta||_p\right)^{-\alpha-n} d^n\eta
$$

$$
= C_1' ||y||_p^{\lambda} \int_{\mathbb{Q}_p^n} \left(1 + ||\tau||_p\right)^{-\alpha-n} d\tau \leq C_1 ||y||_p^{\lambda}.
$$

Proof of Claim C

$$
I_3(y) = \sum_{k=l+1}^{\infty} \int_{||\eta||_p=p^k} \left(1 + ||\eta||_p\right)^{-\alpha-n} ||\eta||_p^{\lambda} d^n\eta,
$$

$$
\leq \int_{\mathbb{Q}_p^n} \left(1 + ||\eta||_p\right)^{-\alpha-n} ||\eta||_p^{\lambda} d^n\eta = C. \qquad \blacksquare
$$

Lemma 46 *The functions u_1, u_2 belong to $\mathfrak{M}_\lambda$ uniformly in t, for $\lambda + n < \alpha$.*

Proof By Lemma 44 (iii) and Proposition 45,

$$
|u_1(x,t)| \leq \int_{\mathbb{Q}_p^n} Z(x-y,t) \, |\varphi(y)| \, d^n y
$$

$$
\leq C \int_{\mathbb{Q}_p^n} t \left(t^{\frac{1}{\alpha-n}} + ||x-y||_p\right)^{-\alpha} \left(1 + ||y||_p^{\lambda}\right) d^n y
$$

$$
\leq C' \left(1 + ||x||_p^{\lambda}\right).
$$

On the other hand, since $u_1(x,t) = \int_{\mathbb{Q}_p^n} Z(w,t)\varphi(x-w)d^n w$, u_1 is locally constant and $l(u_1) = l(\varphi)$ uniformly in t. The proof for u_2 is similar. $\blacksquare$

Remark 47 Notice that $u_1, u_2, W_\gamma u_1, W_\gamma u_2 \in \mathfrak{M}_\lambda$, for any γ satisfying $\lambda + n < \gamma \leq \alpha$.

3.3.2   Claim $u(x,t)$ Satisfies the Initial Condition

This claim follows from Lemma 46 by using the following result.

Lemma 48 *If $\varphi \in \mathfrak{M}_\lambda$, with $\alpha > \lambda + n$, then $\lim_{t\to 0+} \int_{\mathbb{Q}_p^n} Z(x-\xi,t)\varphi(\xi)d^n\xi = \varphi(x)$.*

Proof By Lemma 44 (iv),

$$\int_{\mathbb{Q}_p^n} Z(x-\xi,t)\varphi(\xi)d^n\xi = \int_{\mathbb{Q}_p^n} Z(x-\xi,t)\left[\varphi(\xi)-\varphi(x)\right]d^n\xi + \varphi(x). \tag{3.7}$$

Now, by Lemma 44 (iii) and the local constancy of φ,

$$\int_{\mathbb{Q}_p^n} Z(x-\xi,t)\left[\varphi(\xi)-\varphi(x)\right]d^n\xi$$

$$\leq Ct\int_{\|x-\xi\|_p \geq p^l} (t^{\frac{1}{\alpha-n}} + \|x-\xi\|_p)^{-\alpha}\,|\varphi(\xi)-\varphi(x)|\,d^n\xi$$

$$\leq Ct\int_{\|z\|_p \geq p^l} (t^{\frac{1}{\alpha-n}} + \|z\|_p)^{-\alpha}\,|\varphi(x-z)-\varphi(x)|\,d^n z$$

$$\leq Ct\int_{\|z\|_p \geq p^l} \|z\|_p^{-\alpha}\,(1+\|x-z\|^\lambda)d^n z + C't\,|\varphi(x)| \leq th(x).$$

Now, the formula is obtained by taking limit $t \to 0^+$ in (3.7). ∎

3.3.3   Claim $u(x,t)$ Is a Solution of Cauchy Problem (3.4)

The proof of this claim is a consequence of Corollary 51, Lemmas 52 and 53. Several preliminary results are required.

Lemma 49 *There exist positive constants C_3, C_4 such that $Z(x,t)$ satisfies the following conditions:*

(i) $\frac{\partial Z(x,t)}{\partial t} = -\kappa \int_{\mathbb{Q}_p^n} A_{w_\alpha}(\|\xi\|_p) e^{-\kappa t A_{w_\alpha}(\|\xi\|_p)} \chi_p(x\cdot\xi)d^n\xi$, *for $t > 0$ and $x \in \mathbb{Q}_p^n$;*

(ii) $\left|\frac{\partial Z(x,t)}{\partial t}\right| \leq C_3 t^{-\frac{\alpha}{\alpha-n}}$, *for $t > 0$ and $x \in \mathbb{Q}_p^n$;*

(iii) $\left|\frac{\partial Z(x,t)}{\partial t}\right| \leq C_4 t \|x\|_p^{n-2\alpha}$, *for $t > 0$ and $x \in \mathbb{Q}_p^n\backslash\{0\}$;*

(iv) $\left|\frac{\partial Z(x,t)}{\partial t}\right| \leq 2^\alpha C_3 \left(\|x\|_p + t^{\frac{1}{\alpha-n}}\right)^{-\alpha}$, *for $t > 0$ and $x \in \mathbb{Q}_p^n \backslash \{0\}$.*

Proof (i) The formula is obtained by the Lebesgue dominated convergence theorem, and the fact that $-\kappa A_{w_\alpha}(\|\xi\|_p)e^{-\kappa\tau A_{w_\alpha}(\|\xi\|_p)}\chi_p(x\cdot\xi) \in L^1(\mathbb{Q}_p^n)$, for $\tau > 0$ fixed, cf. Chap. 2, Lemma 8. (ii) By using (i) and Lemma 8 in Chap. 2,

$$\left|\frac{\partial Z(x,t)}{\partial t}\right| \leq \int_{\mathbb{Q}_p^n} C_1 \|\xi\|_p^{\alpha-n} e^{-\kappa C_2 t\|\xi\|_p^{\alpha-n}} d^n\xi.$$

We now pick an integer m such that $p^{m-1} \leq t^{\frac{1}{\alpha-n}} \leq p^m$, and proceed as in the proof of Lemma 44 (i), to obtain

$$\left|\frac{\partial Z(x,t)}{\partial t}\right| \leq C_1 p^{-(m-1)n-(m-1)(\alpha-n)} \int_{\mathbb{Q}_p^n} \|z\|_p^{\alpha-n} e^{-\kappa C_2\|z\|_p^{\alpha-n}} d^n z \leq C_3 t^{-\frac{\alpha}{\alpha-n}}.$$

(iii) Set $\|x\|_p = p^\beta$. Now, since $A_{w_\alpha}(\|\xi\|_p)e^{-\kappa t A_{w_\alpha}(\|\xi\|_p)} \in L^1 \cap L^2$ for $t > 0$, then $\frac{\partial Z(x,t)}{\partial t} \in L^1 \cap L^2$ for $t > 0$, and by applying the formula for the Fourier Transform of a radial function, we get

$$\frac{\partial Z(x,t)}{\partial t} = \|x\|_p^{-n}\left((1-p^{-n})\sum_{j=0}^{\infty} A_{w_\alpha}(p^{-\beta-j})e^{-\kappa t A_{w_\alpha}(p^{-\beta-j})}p^{-nj}\right.$$
$$\left. -A_{w_\alpha}(p^{-\beta+1})e^{-\kappa t A_{w_\alpha}(p^{-\beta+1})}\right).$$

Now, by using that $A_{w_\alpha}(\xi)$ is a decreasing function of $ord(\xi)$,

$$\left|\frac{\partial Z(x,t)}{\partial t}\right| \leq \|x\|_p^{-n} A_{w_\alpha}(p^{-\beta+1})\left|(1-p^{-n})\sum_{j=0}^{\infty}p^{-nj} - e^{-\kappa t A_{w_\alpha}(p^{-\beta+1})}\right|$$
$$\leq \|x\|_p^{-n} A_{w_\alpha}(p^{-\beta+1})\left(1 - e^{-\kappa t A_{w_\alpha}(p^{-\beta+1})}\right)$$

By using mean value theorem and Lemma 8 in Chap. 2, we have

$$\left|\frac{\partial Z(x,t)}{\partial t}\right| \leq C_4 \|x\|_p^{n-2\alpha} t.$$

(iv) If $\|x\|_p \leq t^{\frac{1}{\alpha-n}}$, then $\frac{\|x\|_p}{2} + \frac{t^{\frac{1}{\alpha-n}}}{2} \leq t^{\frac{1}{\alpha-n}}$ and $t^{\frac{-\alpha}{\alpha-n}} \leq 2^\alpha\left(\|x\|_p + t^{\frac{1}{\alpha-n}}\right)^{-\alpha}$, multiplying by C_3 and using (ii), we have

$$\left|\frac{\partial Z(x,t)}{\partial t}\right| \leq 2^\alpha C_3 \left(\|x\|_p + t^{\frac{1}{\alpha-n}}\right)^{-\alpha}.$$

Now, if $\|x\|_p \geq t^{\frac{1}{\alpha-n}}$, by using (iii),

$$\left| \frac{\partial Z(x,t)}{\partial t} \right| \leq C_3 \|x\|_p^{-\alpha}, \tag{3.8}$$

and since $\|x\|_p \geq t^{\frac{1}{\alpha-n}}$, then $\|x\|_p \geq \left(\frac{\|x\|_p}{2} + \frac{t^{\frac{1}{\alpha-n}}}{2} \right)$ and $2^{\alpha} \left(\|x\|_p + t^{\frac{1}{\alpha-n}} \right)^{-\alpha} \geq$
$\|x\|_p^{-\alpha}$, multiplying by C_3 and using (3.8), we have

$$\left| \frac{\partial Z(x,t)}{\partial t} \right| \leq 2^{\alpha} C_3 \left(\|x\|_p + t^{\frac{1}{\alpha-n}} \right)^{-\alpha}. \qquad \blacksquare$$

Lemma 50 $(\mathbf{W}_\gamma Z_t)(x)$, with $\gamma \leq \alpha$, satisfies the following conditions:

(i) $(\mathbf{W}_\gamma Z_t)(x) = -\int_{\mathbb{Q}_p^n} A_{w_\gamma}(\|\xi\|_p) e^{-\kappa t A_{w_\alpha}(\|\xi\|_p)} \chi_p(x \cdot \xi) d^n \xi$, for $t > 0$ and $x \in \mathbb{Q}_p^n$;

(ii) $\left| (\mathbf{W}_\gamma Z_t)(x) \right| \leq 2^\gamma C \left(\|x\|_p + t^{\frac{1}{\alpha-n}} \right)^{-\gamma}$, for $t > 0$ and $x \in \mathbb{Q}_p^n$ and some positive constant C;

(iii) $\int_{\mathbb{Q}_p^n} (\mathbf{W}_\gamma Z_t)(x) d^n x = 0$.

Proof

(i) Define

$$Z_t^{(M)}(x) = \int_{\|\eta\|_p \leq p^M} \chi_p(x \cdot \eta) e^{-\kappa t A_{w_\alpha}(\|\eta\|_p)} d^n \eta, \text{ for } M \in \mathbb{N}. \tag{3.9}$$

This function is locally constant on $\mathbb{Q}_p^n$. Indeed, if $\|\xi\|_p \leq p^{-M}$, then $Z_t^{(M)}(x + \xi) = Z_t^{(M)}(x)$. Furthermore, $Z_t^{(M)}(x)$ is bounded, and thus $Z_t^{(M)}(x) \in \mathfrak{M}_0 \subset Dom(\mathbf{W}_\gamma)$. We now use formula (3.2) and Fubini's theorem to compute $(\mathbf{W}_\gamma Z_t^{(M)})(x)$ as follows:

$$(\mathbf{W}_\gamma Z_t^{(M)})(x) = \int_{\mathbb{Q}_p^n} \frac{Z_t^{(M)}(x - \xi) - Z_t^{(M)}(x)}{w_\gamma(\|\xi\|_p)} d^n \xi$$

$$= \int_{\|\xi\|_p > p^{-M}} \int_{\|\eta\|_p \leq p^M} e^{-\kappa t A_{w_\alpha}(\|\eta\|_p)} \chi_p(x \cdot \eta) \frac{(\chi_p(\xi \cdot \eta) - 1)}{w_\gamma(\|\xi\|_p)} d^n \eta d^n \xi$$

$$= \int_{\|\eta\|_p \leq p^M} e^{-\kappa t A_{w_\alpha}(\|\eta\|_p)} \chi_p(x \cdot \eta) \int_{\|\xi\|_p > p^{-M}} \frac{(\chi_p(\xi \cdot \eta) - 1)}{w_\gamma(\|\xi\|_p)} d^n \xi d^n \eta$$

$$= - \int_{\|\eta\|_p \leq p^M} e^{-\kappa t A_{w_\alpha}(\|\eta\|_p)} \chi_p(x \cdot \eta) A_{w_\gamma}(\|\eta\|_p) d^n \eta.$$

By using that $e^{-\kappa t A_{w_\alpha}(\|\xi\|_p)} A_{w_\gamma}(\|\xi\|_p) \in L^1(\mathbb{Q}_p^n)$ for $t > 0$, and the dominated convergence theorem, we obtain

$$\lim_{M \to \infty} (\mathbf{W}_\gamma Z_t^{(M)})(x) = -\int_{\mathbb{Q}_p^n} A_{w_\gamma}(\|\eta\|_p) e^{-\kappa t A_{w_\alpha}(\|\eta\|_p)} \chi_p(x \cdot \eta) d^n \eta. \qquad (3.10)$$

On the other hand, by fixing $x \neq 0$ and for $t > 0$, $Z_t(x - \xi) - Z_t(x)$ is locally constant, cf. Remark 43, and bounded, cf. Lemma 44 (iii), then $(\mathbf{W}_\gamma Z_t)(x)$ is well-defined, and since $Z_t^{(M)}(x)$ is radial,

$$(\mathbf{W}_\gamma Z_t^{(M)})(x) = \int_{\|\xi\|_p > \|x\|_p} \frac{Z_t^{(M)}(x - \xi) - Z_t^{(M)}(x)}{w_\gamma(\|\xi\|_p)} d^n \xi,$$

and by the dominated convergence theorem, $\lim_{M \to \infty} (\mathbf{W}_\gamma Z_t^{(M)})(x) = (\mathbf{W}_\gamma Z_t)(x)$. Therefore by (3.10), we have

$$(\mathbf{W}_\gamma Z_t)(x) = -\int_{\mathbb{Q}_p^n} A_{w_\gamma}(\|\eta\|_p) e^{-\kappa t A_{w_\alpha}(\|\eta\|_p)} \chi_p(x \cdot \eta) d^n \eta.$$

Finally, we note the right-hand side in the above formula is continuous at $x = 0$.

(ii) By (i) and Lemma 8 in Chap. 2,

$$\left| (\mathbf{W}_\gamma Z_t)(x) \right| \leq C_0 \int_{\mathbb{Q}_p^n} \|\xi\|_p^{\gamma-n} e^{-\kappa C_1 t \|\xi\|_p^{\alpha-n}} d^n \xi.$$

We now pick an integer m such that $p^{m-1} \leq t^{\frac{1}{\alpha-n}} \leq p^m$, and proceed as in the proof of Lemma 44 (i), to obtain

$$\left| (\mathbf{W}_\gamma Z_t)(x) \right| \leq C t^{-\frac{\gamma}{\alpha-n}}. \qquad (3.11)$$

Now, if $\|x\|_p \leq t^{\frac{1}{\alpha-n}}$, then

$$\frac{\|x\|_p}{2} + \frac{t^{\frac{1}{\alpha-n}}}{2} \leq t^{\frac{1}{\alpha-n}} \text{ and } t^{\frac{-\gamma}{\alpha-n}} \leq 2^\gamma \left(\|x\|_p + t^{\frac{1}{\alpha-n}} \right)^{-\gamma},$$

multiplying by C and by using (3.11), we have

$$\left| (\mathbf{W}_\gamma Z_t)(x) \right| \leq 2^\gamma C \left(\|x\|_p + t^{\frac{1}{\alpha-n}} \right)^{-\gamma}.$$

On the other hand, let $\|x\|_p = p^\beta$, since $A_{w_\gamma}(\|\xi\|_p)e^{-\kappa t A_{w_\alpha}(\|\xi\|_p)} \in L^1 \cap L^2$ for $t > 0$, then $(\mathbf{W}_\gamma Z_t)(x) \in L^1 \cap L^2$ for $t > 0$, by proceeding as in the proof of Lemma 49 (iii), we obtain

$$\left|(\mathbf{W}_\gamma Z_t)(x)\right| \le Ct \|x\|_p^{n-\alpha-\gamma}.$$

Now, if $\|x\|_p \ge t^{\frac{1}{\alpha-n}}$, then

$$\left|(\mathbf{W}_\gamma Z_t)(x)\right| \le C \|x\|_p^{-\gamma}. \tag{3.12}$$

If $\|x\|_p \ge t^{\frac{1}{\alpha-n}}$, then $\|x\|_p \ge \left(\frac{\|x\|_p}{2} + \frac{t^{\frac{1}{\alpha-n}}}{2}\right)$ and $2^\gamma \left(\|x\|_p + t^{\frac{1}{\alpha-n}}\right)^{-\gamma} \ge \|x\|_p^{-\gamma}$, multiplying by C and using (3.12), we have

$$\left|(\mathbf{W}_\gamma Z_t)(x)\right| \le 2^\gamma C \left(\|x\|_p + t^{\frac{1}{\alpha-n}}\right)^{-\gamma}.$$

(iii) It follows from (i) by the inversion formula for the Fourier transform. ∎

Corollary 51 $\frac{\partial Z(x,t)}{\partial t} = \kappa \cdot (\mathbf{W}_\alpha Z_t)(x)$ for $t > 0$ and $x \in \mathbb{Q}_p^n$.

Proof The formula follows from Lemma 49 (i) and Lemma 50 (i). ∎

Proposition 52 *Assume that $\varphi \in \mathfrak{M}_\lambda$, then the following assertions hold:*

(i) $\frac{\partial u_1}{\partial t}(x, t) = \int_{\mathbb{Q}_p^n} \frac{\partial Z(x-y,t)}{\partial t}\varphi(y)d^n y$, *for $t > 0$ and $x \in \mathbb{Q}_p^n \backslash \{0\}$;*

(ii) $(\mathbf{W}_\gamma u_1)(x, t) = \int_{\mathbb{Q}_p^n}(\mathbf{W}_\gamma Z_t)(x - y)\varphi(y)d^n y$, *for $n + \lambda < \gamma \le \alpha$, $t > 0$ and $x \in \mathbb{Q}_p^n \backslash \{0\}$.*

Proof

(i) By using the mean value theorem, $\frac{\partial u_1}{\partial t}(x, t)$ equals

$$\lim_{h \to 0} \int_{\mathbb{Q}_p^n} \left[\frac{Z(x - y, t + h) - Z(x - y, t)}{h}\right]\varphi(y)d^n y = \lim_{h \to 0}\int_{\mathbb{Q}_p^n}\frac{\partial Z(x - y, \tau)}{\partial t}\varphi(y)d^n y,$$

where τ is between t and $t + h$. Now, the result follows by applying the dominated convergence theorem and Lemma 49 (iv).

(ii) By Remark 47, if $n + \lambda < \gamma$, then $u_1 \in Dom(\mathbf{W}_\gamma)$ for $t > 0$. Then for any $L \in \mathbb{N}$, by Fubini's theorem, cf. Lemma 44 (iii), we get

$$\int_{\|y\|_p > p^{-L}}\frac{u_1(x - y, t) - u_1(x, t)}{w_\gamma(\|y\|_p)}d^n y = \int_{\mathbb{Q}_p^n}\varphi(\xi)\int_{\|y\|_p > p^{-L}}\frac{(Z_t(x - \xi - y) - Z_t(x - \xi))}{w_\gamma(\|y\|_p)}$$

$$\times d^n y d^n \xi.$$

We now fix a positive integer M, such that $\|y\|_p < p^{-L} < p^{-M} < \|x - \xi\|_p$, by using Remark 43,

$$\int_{\|x-\xi\|_p > p^{-M}} \varphi(\xi) \int_{\|y\|_p \le p^{-L}} \frac{(Z_t(x - \xi - y) - Z_t(x - \xi))}{w_\gamma(\|y\|_p)} d^n y d^n \xi = 0,$$

then

$$(W_\gamma u_1)(x, t) = \lim_{L \to \infty} \int_{\|y\|_p > p^{-L}} \frac{u_1(x - y, t) - u_1(x, t)}{w_\gamma(\|y\|_p)} d^n y$$

$$= \int_{\|x-\xi\|_p > p^{-M}} \varphi(\xi)(W_\gamma Z_t)(x - \xi) d^n \xi$$

$$+ \lim_{L \to \infty} \int_{\|x-\xi\|_p \le p^{-M}} \varphi(\xi) \int_{\|y\|_p > p^{-L}} \frac{(Z_t(x - \xi - y) - Z_t(x - \xi))}{w_\gamma(\|y\|_p)} d^n y d^n \xi$$

$$= \int_{\|x-\xi\|_p > p^{-M}} \varphi(\xi)(W_\gamma Z_t)(x - \xi) d^n \xi$$

$$+ \int_{\|x-\xi\|_p \le p^{-M}} \varphi(\xi)(W_\gamma Z_t)(x - \xi) d^n \xi,$$

where the limit was computed by using the Lebesgue dominated convergence theorem and the fact that

$$\int_{\|x-\xi\|_p \le p^{-M}} \varphi(\xi) \int_{\|y\|_p > p^{-L}} \frac{(Z_t(x - \xi - y) - Z_t(x - \xi))}{w_\gamma(\|y\|_p)} d^n y d^n \xi$$

$$= \int_{\|x-\xi\|_p \le p^{-M}} \varphi(\xi) \int_{\|y\|_p > p^{-M}} \frac{(Z_t(x - \xi - y) - Z_t(x - \xi))}{w_\gamma(\|y\|_p)} d^n y d^n \xi,$$

because $Z_t(x - \xi - y) = Z_t(x - \xi)$ for $\|x - \xi\|_p > \|y\|_p$, cf. Remark 43. The convergence of this last integral follows from Proposition 45.

∎

Set $u_2(x, t, \tau) := \int_\tau^t \int_{\mathbb{Q}_p^n} Z(x - y, t - \theta) f(y, \theta) d^n y d\theta$, for $f \in \mathfrak{M}_\lambda$ with $\alpha - n > \lambda$, for $0 \le \tau \le t \le T$, and $x \in \mathbb{Q}_p^n$. By reasoning as in the proof of Lemma 46, we have $u_2(x, t, \tau) \in \mathfrak{M}_\lambda$ uniformly in t and τ.

Proposition 53 *Assume that* $f \in \mathfrak{M}_\lambda$, *with* $\alpha - n > \lambda$, *then the following assertions hold:*

(i) $\frac{\partial u_2}{\partial t}(x, t, \tau) = f(x, t) + \int_\tau^t \left(\int_{\mathbb{Q}_p^n} \frac{\partial Z(x-y,t-\theta)}{\partial t} [f(y, \theta) - f(x, \theta)] \, d^n y \right) d\theta$, *for* $t > 0$
and $x \in \mathbb{Q}_p^n$;

(ii) $(\mathbf{W}_\gamma u_2)(x, t, \tau) = \int_\tau^t \int_{\mathbb{Q}_p^n} (\mathbf{W}_\gamma Z)(x - y, t - \theta) f(y, \theta) d^n y d\theta$, *for* $n + \lambda < \gamma \le \alpha$,
$t > 0$ *and* $x \in \mathbb{Q}_p^n$.

Proof Set

$$
u_{2,h}(x, t, \tau) := \int_\tau^{t-h} \int_{\mathbb{Q}_p^n} Z(x - y, t - \theta) f(y, \theta) d^n y d\theta, \ 0 < h < t - \tau.
$$

By using a standard reasoning, one shows that

$$
\frac{\partial u_{2,h}}{\partial t}(x, t, \tau) = \int_\tau^{t-h} \int_{\mathbb{Q}_p^n} \frac{\partial Z(x - y, t - \theta)}{\partial t} f(y, \theta) d^n y d\theta + \int_{\mathbb{Q}_p^n} Z(x-y, h) f(y, t-h) d^n y d\theta.
$$

This formula can be rewritten as

$$
\frac{\partial u_{2,h}}{\partial t}(x, t, \tau) = \int_\tau^{t-h} \int_{\mathbb{Q}_p^n} \frac{\partial Z(x - y, t - \theta)}{\partial t} [f(y, \theta) - f(x, \theta)] d^n y d\theta
$$

$$
+ \int_\tau^{t-h} f(x, \theta) \int_{\mathbb{Q}_p^n} \frac{\partial Z(x - y, t - \theta)}{\partial t} d^n y d\theta
$$

$$
+ \int_{\mathbb{Q}_p^n} Z(x - y, h) [f(y, t - h) - f(y, t)] d^n y
$$

$$
+ \int_{\mathbb{Q}_p^n} Z(x - y, h) f(y, t) d^n y.
$$

The first integral contains no singularity at $t = \theta$ due to Lemma 49 (iv) and the local constancy of f. By Lemma 44 (iv), the second integral is equal to zero. The third integral can be written as a sum of the integrals over $\{y \in \mathbb{Q}_p^n : \|x - y\|_p \ge p^M\}$ and the complement of this set, one of these integrals is estimated on the basis of the uniform continuity of f, while the other contains no singularity, see Lemma 49 (iv). Finally, the fourth integral tends to $f(x, t)$ as $h \to 0^+$, cf. Lemma 48.

(ii) By Lemma 46, $\mathbf{W}_\gamma u_{2,h}$ is well-defined for any γ satisfying $n + \lambda < \gamma \leq \alpha$. Then, for any $L \in \mathbb{N}$, the following integral exists:

$$
\int\limits_{\|\xi\|_p \geq p^{-L}} \frac{[u_{2,h}(x - \xi, t, \tau) - u_{2,h}(x, t, \tau)]}{w_\gamma(\|\xi\|_p)} d^n \xi
$$

$$
= \int\limits_\tau^{t-h} \int\limits_{\mathbb{Q}_p^n} \int\limits_{\|\xi\|_p \geq p^{-L}} \frac{[Z(x - \xi - y, t - \theta) - Z(x - y, t - \theta)]}{w_\gamma(\|\xi\|_p)} f(y, \theta) d^n \xi \, d^n y \, d\theta.
$$

$$(3.13)$$

On the other hand, by Fubini's theorem,

$$
\int\limits_{\|\xi\|_p \geq p^{-L}} \frac{[Z(x - \xi - y, t - \theta) - Z(x - y, t - \theta)]}{w_\gamma(\|\xi\|_p)} d^n \xi
$$

$$
= \int\limits_{\mathbb{Q}_p^n} \chi_p((x - y) \cdot \eta) e^{-\kappa(t-\theta) A_{w_\gamma}(\|\eta\|_p)} P_k(\eta) d^n \eta,
$$

where $P_k(\eta) = \int_{\|\xi\|_p \geq p^{-L}} \frac{[\chi_p(-\xi \cdot \eta) - 1]}{w_\gamma(\|\xi\|_p)} d^n \xi$. A simple calculation shows that $|P_k(\eta)| \leq C' \|\eta\|_p^{\gamma - n}$, and then

$$
\int\limits_{\|\xi\|_p \geq p^{-L}} \frac{[Z(x - \xi - y, t - \theta) - Z(x - y, t - \theta)]}{w_\gamma(\|\xi\|_p)} d^n \xi \leq C,
$$

where the constant does not depend on $x, t \geq h + \tau, L$. Now, by expressing the right integral of (3.13) as

$$
\int\limits_\tau^{t-h} \int\limits_{\|x - \xi\|_p > p^{-M}} \int\limits_{\|\xi\|_p \geq p^{-L}} \frac{[Z(x - \xi - y, t - \theta) - Z(x - y, t - \theta)]}{w_\gamma(\|\xi\|_p)} f(y, \theta) d^n \xi \, d^n y \, d\theta
$$

$$
+ \int\limits_\tau^{t-h} \int\limits_{\|x - \xi\|_p \leq p^{-M}} \int\limits_{\|\xi\|_p \geq p^{-L}} \frac{[Z(x - \xi - y, t - \theta) - Z(x - y, t - \theta)]}{w_\gamma(\|\xi\|_p)} f(y, \theta) d^n \xi \, d^n y \, d\theta
$$

where M is a positive integer, such that $\|\xi\|_p < p^{-L} < p^{-M} < \|x - \xi\|_p$, and using the same reasoning as in the final part of the proof of Proposition 52 (ii), we obtain

$$(W_\gamma u_{2,h})(x, t) = \int_\tau^{t-h} \int_{\mathbb{Q}_p^n} (W_\gamma Z)(x - \xi, t - \theta) f(y, \theta) d^n \xi d\theta. \tag{3.14}$$

Now, by Lemma 50 (ii), the fact that $f \in \mathfrak{M}_\lambda$, Proposition 45, and the dominated convergence theorem, we can take limit as $h \to 0^+$, which completes the proof when $\gamma < \alpha$. If $\gamma = \alpha$, formula (3.14) remains valid. By using Lemma 50 (iii), formula (3.14) can be rewritten as

$$(W_\gamma u_{2,h})(x, t) = \int_\tau^{t-h} \int_{\mathbb{Q}_p^n} (W_\gamma Z)(x - \xi, t - \theta) \left[f(y, \theta) - f(x, \theta) \right] d^n \xi d\theta.$$

Now, by using the local constancy of f, we can justify the passage to the limit as $h \to 0^+$, which completes the proof. ∎

Remark 54 By Lemma 44 (iv) and Lemma 49 (i), $\int_{\mathbb{Q}_p^n} \frac{\partial Z(x-y, t-\theta)}{\partial t} d^n y = 0$, then

$$\frac{\partial u_2}{\partial t}(x, t, \tau) = f(x, t) + \int_\tau^t \left(\int_{\mathbb{Q}_p^n} \frac{\partial Z(x - y, t - \theta)}{\partial t} f(y, \theta) d^n y \right) d\theta,$$

for $t > 0$ and $x \in \mathbb{Q}_p^n$.

3.4   Parabolic-Type Equations with Variable Coefficients

First, we fix the notation that will be used through this section. We fix $N + 1$ positive real numbers satisfying $n < \alpha_1 < \alpha_2 < \cdots < \alpha_N < \alpha$. We assume that $\alpha > n + 1$. We fix $N + 2$ functions $a_k(x, t)$, $k = 0, \ldots N$ and $b(x, t)$ from $\mathbb{Q}_p^n \times [0, T]$ to $\mathbb{R}$, here T is a positive constant. We assume that: (i) $b(x, t)$ and $a_k(x, t)$, for $k = 0, \ldots, N$, belong (with respect to x) to $\mathfrak{M}_0$ uniformly with respect to $t \in [0, T]$; (ii) $a_0(x, t)$ satisfies the Hölder condition in t with exponent $v \in (0, 1)$ uniformly in x. We also assume the uniform parabolicity condition $a_0(x, t) \geq \mu > 0$ and that $\alpha_{N+1} := n + (\alpha - n)(1 - v) > \alpha_N$. Notice that $\alpha_{N+1} < \alpha$.

Set $\tilde{W} := \sum_{k=1}^N a_k(x, t) W_{\alpha_k} - b(x, t) I$ with domain $\mathfrak{M}_\lambda$, and $0 \leq \lambda + n < \alpha_1$. Notice that $\tilde{W} : \mathfrak{M}_\lambda \to \mathfrak{M}_\lambda$.

In this section we construct a solution for the following initial value problem:

$$\begin{cases} \frac{\partial u}{\partial t}(x,t) - a_0(x,t)(\mathbf{W}_\alpha u)(x,t) - (\tilde{\mathbf{W}} u)(x,t) = f(x,t) \\ u(x,0) = \varphi(x), \end{cases} \tag{3.15}$$

where $x \in \mathbb{Q}_p^n$, $t \in (0,T]$, $\varphi(x) \in \mathfrak{M}_\lambda$, $f(x,t) \in \mathfrak{M}_\lambda$ uniformly with respect to t, with $0 \le \lambda < \alpha_1 - n$, and $f(x,t)$ is continuous in (x,t) (if $a_1(x,t) = \cdots = a_N(x,t) \equiv 0$ then we shall assume that $0 \le \lambda < \alpha - n$).

3.4.1 Parametrized Cauchy Problem

We first study the following Cauchy problem:

$$\begin{cases} \frac{\partial u}{\partial t}(x,t) - a_0(y,\theta)(\mathbf{W}_\alpha u)(x,t) = 0, \; x \in \mathbb{Q}_p^n, t \in (0,T] \\ u(x,0) = \varphi(x), \end{cases} \tag{3.16}$$

where $y \in \mathbb{Q}_p^n$, $\theta > 0$ are parameters. By taking $\kappa = a_0(y,\theta) \ge \mu > 0$ and applying the results of Sect. 3.3, Cauchy problem (3.16) has a fundamental solution given by

$$Z(x,t;y,\theta,w_\alpha,\kappa) := Z(x,t;y,\theta) = \int_{\mathbb{Q}_p^n} \chi_p(x \cdot \xi) e^{-a_0(y,\theta)tA_{w_\alpha}(\|\xi\|_p)} d^n\xi,$$

for $t > 0$ and $x \in \mathbb{Q}_p^n$.

Remark 55 All statements from the Lemmas 44, 49, 50 hold for $Z(x,t;y,\theta)$ and the involved constants do not depend of y and θ. Thus, we have the following estimates:

$$Z(x,t;y,\theta) \le C_1 t \left(\|x\|_p + t^{\frac{1}{\alpha-n}} \right)^{-\alpha}, \text{ for } t > 0; \tag{3.17}$$

$$\left| \frac{\partial Z(x,t;y,\theta)}{\partial t} \right| \le C_2 \left(\|x\|_p + t^{\frac{1}{\alpha-n}} \right)^{-\alpha}, \text{ for } t > 0; \tag{3.18}$$

$$\left| (W_\gamma Z)(x,t;y,\theta) \right| \le C_3 \left(\|x\|_p + t^{\frac{1}{\alpha-n}} \right)^{-\gamma}, \text{ for } t > 0 \text{ and } \gamma \le \alpha. \tag{3.19}$$

And the identities:

$$\int_{\mathbb{Q}_p^n} Z(x,t;y,\theta) d^n x = 1, \text{ for } t > 0; \tag{3.20}$$

$$\frac{\partial Z(x,t;y,\theta)}{\partial t} = -a_0(y,\theta) \int_{\mathbb{Q}_p^n} A_{w_\alpha}(\|\xi\|_p) e^{-a_0(y,\theta)tA_{w_\alpha}(\|\xi\|_p)} \chi_p(x \cdot \xi) d^n\xi, \tag{3.21}$$

for $t > 0$;

$$\left(W_\gamma Z\right)(x, t; y, \theta) = -\int_{\mathbb{Q}_p^n} A_{w_\gamma}(\|\xi\|_p) e^{-a_0(y,\theta)tA_{w_\alpha}(\|\xi\|_p)} \chi_p(x \cdot \xi) d^n\xi, \qquad (3.22)$$

for $t > 0$ and $\gamma \le \alpha$;

$$\int_{\mathbb{Q}_p^n} \left(W_\gamma Z_t\right)(x, t; y, \theta) d^n x = 0. \qquad (3.23)$$

Lemma 56 *There exists a positive constant C, such that*

$$\left| \int_{\mathbb{Q}_p^n} \frac{\partial Z(x - y, t; y, \theta)}{\partial t} d^n y \right| \le C. \qquad (3.24)$$

Proof It follows from (3.20), by using (3.21) and (3.18), that

$$\int_{\mathbb{Q}_p^n} \frac{\partial Z(x, t; y, \theta)}{\partial t} d^n x = 0.$$

Therefore

$$\int_{\mathbb{Q}_p^n} \frac{\partial Z(x - y, t; y, \theta)}{\partial t} d^n y = \int_{\mathbb{Q}_p^n} \left[\frac{\partial Z(x - y, t; y, \theta)}{\partial t} - \frac{\partial Z(x - y, t; x, \theta)}{\partial t} \right] d^n y. \qquad (3.25)$$

The function $\frac{\partial Z(x,t;y,\theta)}{\partial t}$ belongs to $\mathfrak{M}_0$ in the variable y, and its exponent of local constancy is equal to the one for $a_0(y, \theta)$ by (3.21). Then the integral in the right side of (3.25) is actually taken over the set

$$\left\{ y \in \mathbb{Q}_p^n; \|y - x\|_p \ge p^{-l} \right\},$$

where l does not depend on x, t, θ. Now, the announced inequality follows from (3.18). ∎

3.4.2   Heat Potentials

We define *the parameterized heat potentials* as follows:

$$u(x, t, \tau) := \int_\tau^t \int_{\mathbb{Q}_p^n} Z(x - y, t - \theta; y, \theta) f(y, \theta) d^n y d\theta,$$

where $f \in \mathfrak{M}_\lambda$, $0 \leq \lambda < \alpha - n$, f continuous in (y, θ). By using the same argument given to prove Lemma 46, one proves that $u \in \mathfrak{M}_\lambda$ uniformly in t and τ.

We now calculate the derivative with respect to t and the action of the operator $\mathbf{W}_\gamma$ on $u(x, t, \tau)$ for $n + \lambda < \gamma \leq \alpha$.

Proposition 57 *Assume that $f \in \mathfrak{M}_\lambda$, $0 \leq \lambda < \alpha - n$, f continuous in (y, θ). Then the following assertions hold:*

(i) $\frac{\partial u(x,t,\tau)}{\partial t} = f(x, t) + \int_\tau^t \int_{\mathbb{Q}_p^n} \frac{\partial Z(x-y,t-\theta;y,\theta)}{\partial t} f(y, \theta) d^n y d\theta$;

(ii) $(\mathbf{W}_\gamma u)(x, t, \tau) = \int_\tau^t \int_{\mathbb{Q}_p^n} (\mathbf{W}_\gamma Z)(x - y, t - \theta; y, \theta) f(y, \theta) d^n y d\theta$, $\gamma \leq \alpha$.

Proof It is a simple variation of the proof given for Proposition 53. ∎

The following technical result will be used later on.

Lemma 58 *Let*

$$
J(x, \xi, t, \tau) = \int_\tau^t (t - \theta)^{-\rho/\beta} (\theta - \tau)^{-\sigma/\beta} \left(\int_{\mathbb{Q}_p^n} \left[(t - \theta)^{1/\beta} + \|x - \eta\|_p \right]^{-n-b_1} \right.
$$

$$
\left. \times \left[(\theta - \tau)^{1/\beta} + \|\eta - \xi\|_p \right]^{-n-b_2} d^n \eta \right) d\theta,   \tag{3.26}
$$

where $x, \xi \in \mathbb{Q}_p^n$, $0 \leq \tau < t$, $b_1, b_2 > 0$, $\rho + b_1 < \beta$, $\sigma + b_2 < \beta$, and $\beta > 1$. Then

$$
J(x, \xi, t, \tau) \leq C \left\{ B \left(1 - \frac{\rho}{\beta}, 1 - \frac{\sigma + b_2}{\beta} \right) \left((t - \tau)^{1/\beta} + \|x - \xi\|_p \right)^{-n-b_1} \right.
$$

$$
\times (t - \tau)^{-\frac{(\rho+\sigma+b_2-\beta)}{\beta}} + B \left(1 - \frac{\rho + b_1}{\beta}, 1 - \frac{\sigma}{\beta} \right)
$$

$$
\left. \times \left((t - \tau)^{1/\beta} + \|x - \xi\|_p \right)^{-n-b_2} (t - \tau)^{-\frac{(\rho+\sigma+b_1-\beta)}{\beta}} \right\},
$$

where C is a positive constant depends only on b_1, b_2 and $B(\cdot, \cdot)$ denotes the Archimedean Beta function.

Proof We decompose the domain of integration in (3.26) as $A_1 \bigsqcup A_2$, where $A_1 := \left[\tau, \frac{t+\tau}{2} \right) \times \mathbb{Q}_p^n$ and $A_2 := \left[\frac{t+\tau}{2}, t \right] \times \mathbb{Q}_p^n$. We also define

$$
A_{11} = \left\{ (\theta, \eta) \in A_1; (\theta - \tau)^{1/\beta} + \|\eta - \xi\|_p \leq \frac{1}{2} \left[(t - \tau)^{1/\beta} + \|x - \xi\|_p \right] \right\},
$$

$$
A_{12} = A_1 \smallsetminus A_{11},
$$

$$
A_{21} = \left\{ (\theta, \eta) \in A_2; (t - \theta)^{1/\beta} + \|x - \eta\|_p \leq \frac{1}{2} \left[(t - \tau)^{1/\beta} + \|x - \xi\|_p \right] \right\},
$$

$$
A_{22} = A_2 \smallsetminus A_{21}.
$$

We now take $[\tau, t] \times \mathbb{Q}_p^n = \bigsqcup_{j,k=1}^2 A_{jk}$ and $J(x, \xi, t, \tau) = \sum_{j,k=1}^2 J_{j,k}$, where $J_{j,k}$ is the integral over A_{jk}. We use the inequality $(u - v)^\gamma \geq u^\gamma - v^\gamma$, where $0 \leq v \leq \frac{u}{2}$, $\gamma \leq 1$. If $(\theta, \eta) \in A_{11}$, by using that $\beta > 1$, and $\theta - \tau \leq \frac{t-\tau}{2}$,

$$
\begin{aligned}
(t - \theta)^{1/\beta} + \|x - \eta\|_p &= [(t - \tau) - (\theta - \tau)]^{1/\beta} + \|(x - \xi) - (\eta - \xi)\|_p \\
&\geq (t - \tau)^{1/\beta} + \|x - \xi\|_p - \left[(\theta - \tau)^{1/\beta} + \|\eta - \xi\|_p\right] \\
&\geq \frac{1}{2}\left[(t - \tau)^{1/\beta} + \|x - \xi\|_p\right].
\end{aligned}
$$

Then, by using Proposition 45,

$$
\begin{aligned}
J_{11} &\leq 2^{n+b_1}\left[(t - \tau)^{1/\beta} + \|x - \xi\|_p\right]^{-n-b_1} \int_\tau^{\frac{t+\tau}{2}} (t - \theta)^{-\rho/\beta}(\theta - \tau)^{-\sigma/\beta} \\
&\quad \times \left(\int_{\mathbb{Q}_p^n}\left[(\theta - \tau)^{1/\beta} + \|\eta\|_p\right]^{-n-b_2} d^n\eta\right) d\theta \\
&\leq 2^{n+b_1}C\left[(t - \tau)^{1/\beta} + \|x - \xi\|_p\right]^{-n-b_1} \int_\tau^{\frac{t+\tau}{2}} (t - \theta)^{-\rho/\beta}(\theta - \tau)^{-\frac{\sigma+b_2}{\beta}} d\theta.
\end{aligned}
$$

By changing variables as $t - \theta = (t - \tau)\,v$, we get

$$
\begin{aligned}
J_{11} &\leq C(n, b_1)(t - \tau)^{-\frac{\rho+\sigma+b_2}{\beta}}\left[(t - \tau)^{1/\beta} + \|x - \xi\|_p\right]^{-n-b_1} \int_{\frac{1}{2}}^1 v^{-\frac{\rho}{\beta}}(1 - v)^{-\frac{\sigma+b_2}{\beta}} dv \\
&\leq C(n, b_1) B\left(1 - \frac{\rho}{\beta}, 1 - \frac{\sigma + b_2}{\beta}\right)(t - \tau)^{-\frac{(\rho+\sigma+b_2-\beta)}{\beta}}\left[(t - \tau)^{1/\beta} + \|x - \xi\|_p\right]^{-n-b_1}.
\end{aligned}
$$

By a similar reasoning, we get

$$
J_{12} \leq C(n, b_2) B\left(1 - \frac{\rho + b_1}{\beta}, 1 - \frac{\sigma}{\beta}\right)(t - \tau)^{-\frac{(\rho+\sigma+b_1-\beta)}{\beta}}\left[(t - \tau)^{1/\beta} + \|x - \xi\|_p\right]^{-n-b_2}.
$$

Similar estimates are valid for J_{21} and J_{22}, by combining them with the estimates for J_{11} and J_{12}, we obtain inequality (3.26). ∎

3.4.3 Construction of a Solution

Theorem 59 *The Cauchy problem (2.16), has a solution, which can be represented in the form*

$$u(x, t) = \int_0^t \int_{\mathbb{Q}_p^n} \Lambda(x, t, \xi, \tau) f(\xi, \tau) d^n\xi d\tau + \int_{\mathbb{Q}_p^n} \Lambda(x, t, \xi, 0) \varphi(\xi) d^n\xi, \qquad (3.27)$$

where the fundamental solution $\Lambda(x, t, \xi, \tau)$, $x, \xi \in \mathbb{Q}_p^n$, $0 \le \tau < t \le T$, has the form

$$\Lambda(x, t, \xi, \tau) = Z(x - \xi, t - \tau; \xi, \tau) + \mathcal{W}(x, t, \xi, \tau), \qquad (3.28)$$

with

$$|\mathcal{W}(x, t, \xi, \tau)| \le C \left\{ (t - \tau)^{2 - \frac{\lambda}{\alpha - n}} \left[(t - \tau)^{\frac{1}{\alpha - n}} + \|x - \xi\|_p \right]^{-\alpha} \right.$$

$$\left. + (t - \tau) \sum_{k=1}^{N+1} \left[(t - \tau)^{\frac{1}{\alpha - n}} + \|x - \xi\|_p \right]^{-\alpha_k} \right\}. \qquad (3.29)$$

Furthermore $Z(x, t; y, \theta)$ satisfies the estimates (3.17), (3.18), (3.19), (3.24).

Proof We use the usual parametrix method, see e.g. [45, 80]. Our proof is essentially self-contained. We look for a fundamental solution of (3.15) having form (3.28), with

$$\mathcal{W}(x, t, \xi, \tau) = \int_\tau^t \int_{\mathbb{Q}_p^n} Z(x - \eta, t - \theta; \eta, \theta) \Phi(\eta, \theta, \xi, \tau) d^n\eta d\theta,$$

and satisfying

$$\frac{\partial \Lambda}{\partial t}(x, t, \xi, \tau) - a_0(x, t)(\mathbf{W}_\alpha \Lambda)(x, t, \xi, \tau) - \sum_{k=1}^{N} a_k(x, t)(\mathbf{W}_{\alpha_k} \Lambda)(x, t, \xi, \tau)$$

$$+ b(x, t)\Lambda(x, t, \xi, \tau) = 0, \qquad (3.30)$$

for $x \neq 0$, $t > 0$. Now by using (3.28), (3.21)–(3.22) and Proposition 57, we have formally

$$\frac{\partial Z}{\partial t}(x - \xi, t - \tau, \xi, \tau) + \Phi(x, t, \xi, \tau) + \int_{\tau}^{t} \int_{\mathbb{Q}_p^n} \frac{\partial Z(x - \eta, t - \theta; \eta, \theta)}{\partial t} \Phi(\eta, \theta, \xi, \tau) d^n \eta d\theta$$

$$- a_0(x, t) \Big\{ (\mathbf{W}_\alpha Z)(x - \xi, t - \tau; \xi, \tau) + \int_{\tau}^{t} \int_{\mathbb{Q}_p^n} (\mathbf{W}_\alpha Z)(x - \eta, t - \theta; \eta, \theta)$$

$$\times \Phi(\eta, \theta, \xi, \tau) d^n \eta d\theta \Big\} - \sum_{k=1}^{N} a_k(x, t) \Big\{ (\mathbf{W}_{\alpha_k} Z)(x - \xi, t - \tau; \xi, \tau)$$

$$+ \int_{\tau}^{t} \int_{\mathbb{Q}_p^n} (\mathbf{W}_{\alpha_k} Z)(x - \eta, t - \theta; \eta, \theta) \Phi(\eta, \theta, \xi, \tau) d^n \eta d\theta \Big\}$$

$$+ b(x, t) \Big\{ Z(x - \xi, t - \tau; \xi, \tau) + \int_{\tau}^{t} \int_{\mathbb{Q}_p^n} Z(x - \eta, t - \theta; \eta, \theta) \Phi(\eta, \theta, \xi, \tau) d^n \eta d\theta \Big\} = 0.$$

By taking

$$R(x, t, \xi, \tau) := (a_0(x, t) - a_0(\xi, \tau))(\mathbf{W}_\alpha Z)(x - \xi, t - \tau; \xi, \tau)$$

$$+ \sum_{k=1}^{N} a_k(x, t)(\mathbf{W}_{\alpha_k} Z)(x - \xi, t - \tau; \xi, \tau) - b(x, t) Z(x - \xi, t - \tau; \xi, \tau),$$

one gets that $\Phi(x, t, \xi, \tau)$ satisfies the integral equation

$$\Phi(x, t, \xi, \tau) = R(x, t, \xi, \tau) + \int_{\tau}^{t} \int_{\mathbb{Q}_p^n} R(x, t, \eta, \theta) \Phi(\eta, \theta, \xi, \tau) d^n \eta d\theta. \qquad (3.31)$$

Now, by using (3.19) and (3.17), we obtain

$$|R(x, t, \xi, \tau)| \leq C_0 \Big(|a_0(x, t) - a_0(\xi, \tau)| \big((t - \tau)^{\frac{1}{\alpha - n}} + \|x - \xi\|_p \big)^{-\alpha}$$

$$+ \sum_{k=1}^{N} \big((t - \tau)^{\frac{1}{\alpha - n}} + \|x - \xi\|_p \big)^{-\alpha_k}$$

$$+ \big((t - \tau)^{\frac{1}{\alpha - n}} + \|x - \xi\|_p \big)^{-\alpha} (t - \tau) \Big). \qquad (3.32)$$

Claim A

$$|a_0(x,t) - a_0(\xi,\tau)| \left((t-\tau)^{\frac{1}{\alpha-n}} + \|x-\xi\|_p\right)^{-\alpha} \le C_1'\left((t-\tau)^{\frac{1}{\alpha-n}} + \|x-\xi\|_p\right)^{-\alpha_{N+1}},$$

where $\alpha_{N+1} = n + (\alpha - n)(1 - v) > \alpha_N$.

The proof is based on the Hölder condition for $a_0(x,t)$:

$$|a_0(x,t) - a_0(\xi,\tau)| \le C_1 (t-\tau)^v + |a_0(x,\tau) - a_0(\xi,\tau)|.$$

Let $l(a_0)$ be the parameter of local constancy of a_0. Thus, if $\|x - \xi\|_p \le p^{l(a_0)}$, then $|a_0(x,t) - a_0(\xi,\tau)| \le C_1 (t-\tau)^v$. In the case $\|x - \xi\|_p \le p^{l(a_0)}$, the inequality follows from the fact that $(t-\tau)^v \left((t-\tau)^{\frac{1}{\alpha-n}} + \|x-\xi\|_p\right)^{-\alpha+\alpha_{N+1}}$ is bounded, which in turn follows from $\lim_{t \to \tau} (t-\tau)^{v + \frac{-\alpha+\alpha_{N+1}}{\alpha-n}} = 1$. In the case $\|x-\xi\|_p > p^{l(a_0)}$, by using that $|a_0(x,t) - a_0(\xi,\tau)| \le C_0$, the inequality follows from

$$\left((t-\tau)^{\frac{1}{\alpha-n}} + \|x-\xi\|_p\right)^{-\alpha+\alpha_{N+1}} \le \|x-\xi\|_p^{-\alpha+\alpha_{N+1}} \le p^{(-\alpha+\alpha_{N+1})l(a_0)}.$$

Claim B

$$(t-\tau)\left((t-\tau)^{\frac{1}{\alpha-n}} + \|x-\xi\|_p\right)^{-\alpha} \le C_2\left((t-\tau)^{\frac{1}{\alpha-n}} + \|x-\xi\|_p\right)^{-\alpha_{N+1}}.$$

This assertion is a consequence of the fact that $\lim_{t \to \tau} (t-\tau)^{1 + \frac{-\alpha+\alpha_{N+1}}{\alpha-n}} = 0$.

Now from (3.32), and Claims A–B, we have

$$|R(x,t,\xi,\tau)| \le C \sum_{k=1}^{N+1} \left[(t-\tau)^{\frac{1}{\alpha-n}} + \|x-\xi\|_p\right]^{-\alpha_k}. \tag{3.33}$$

We solve integral equation (3.31) by the method of successive approximations:

$$\Phi(x,t,\xi,\tau) = \sum_{m=1}^{\infty} R_m(x,t,\eta,\theta), \tag{3.34}$$

where $R_1 \equiv R$ and

$$R_{m+1}(x,t,\xi,\tau) = \int_{\tau}^{t} \int_{\mathbb{Q}_p^n} R(x,t,\eta,\theta) R_m(\eta,\theta,\xi,\tau) d^n\eta d\theta, \text{ for } m \ge 1.$$

Claim C

$$|R_{m+1}(x, t, \xi, \tau)| \le C(2N+2)^m (t-\tau)^{mv} \frac{(\Gamma(v))^{m+1}}{\Gamma((m+1)v)}$$

$$\times \sum_{j=1}^{N+1} \left[(t-\tau)^{\frac{1}{\alpha-n}} + \|x-\xi\|_p \right]^{-\alpha_j},$$

for $m \ge 0$, where $\Gamma(\cdot)$ denotes the Archimedean Gamma function.

The proof of this assertion will be given later.

It follows from Claim A, by the Stirling formula, that series (3.34) is convergent and that

$$|\Phi(x, t, \xi, \tau)| \le C_0 \sum_{k=1}^{N+1} \left[(t-\tau)^{\frac{1}{\alpha-n}} + \|x-\xi\|_p \right]^{-\alpha_k}. \tag{3.35}$$

Now (3.29) follows from (3.35) and Lemma 58.

Denote by $u_1(x, t)$ and $u_2(x, t)$ the first and second terms in the right hand side of (3.27). Substituting (3.28) into (3.27), we find that

$$u_1(x, t) = \int_0^t \int_{\mathbb{Q}_p^n} Z(x-\xi, t-\tau; \xi, \tau) f(\xi, \tau) d^n \xi d\tau$$

$$+ \int_0^t \int_{\mathbb{Q}_p^n} Z(x-\eta, t-\theta; \eta, \theta) F(\eta, \theta) d^n \eta d\theta,$$

and

$$u_2(x, t) = \int_{\mathbb{Q}_p^n} Z(x-\xi, t; \xi, 0) \varphi(\xi) d^n \xi + \int_0^t \int_{\mathbb{Q}_p^n} Z(x-\eta, t-\theta; \eta, \theta) G(\eta, \theta) d^n \eta d\theta,$$

where

$$F(\eta, \theta) = \int_0^\theta \int_{\mathbb{Q}_p^n} \Phi(\eta, \theta, \xi, \tau) f(\xi, \tau) d^n \xi d\tau, \tag{3.36}$$

$$G(\eta, \theta) = \int_{\mathbb{Q}_p^n} \Phi(\eta, \theta, \xi, 0) \varphi(\xi) d^n \xi. \tag{3.37}$$

Now, by Proposition 45 and (3.35), it follows that

$$|F(\eta, \theta)| \le C_0(1 + \|\eta\|_p^\lambda), \quad |G(\eta, \theta)| \le C_1(1 + \|\eta\|_p^\lambda),$$

for all $\eta \in \mathbb{Q}_p^n$ and $\theta \in (0, T]$.

Claim D The functions F and G belong to $\tilde{\mathcal{E}}$, and their parameters of local constancy do not depend on θ.

We first note that by (3.36)–(3.37), it is sufficient to show that $\Phi(\cdot, \theta, \star, \tau)$ is a locally constant function on $\left(\mathbb{Q}_p^\times\right)^n \times \mathbb{Q}_p^n$ and that its parameter of local constancy do not depend on θ and τ. Now, by the recursive definition of the function Φ we see that if L is the parameter local constancy for all the functions $a_k(\cdot, t)$, $b(\cdot, t)$, $(\mathbf{W}_{\alpha_k}Z)(\cdot, t - \tau; \star, \tau)$ and $Z(\cdot, t - \tau; \star, \tau)$ on $\left(\mathbb{Q}_p^\times\right)^n \times \mathbb{Q}_p^n$, and if $\|\delta\|_p \le p^{-L}$, we have $R(x + \delta, t, \xi + \delta, \tau) = R(x, t, \xi, \tau)$. Furthermore, we successively obtain

$$R_{m+1}(x + \delta, t, \xi + \delta, \tau) = \int_\tau^t \int_{\mathbb{Q}_p^n} R(x + \delta, t, \eta, \theta) R_m(\eta, \theta, \xi + \delta, \tau) d^n\eta \, d\theta$$

$$= \int_\tau^t \int_{\mathbb{Q}_p^n} R(x + \delta, t, \zeta + \delta, \theta) R_m(\zeta + \delta, \theta, \xi + \delta, \tau) d^n\zeta \, d\theta$$

$$= R_{m+1}(x, t, \xi, \tau),$$

so that $\Phi(x + \delta, t, \xi + \delta, \tau) = \Phi(x, t, \xi, \tau)$, and hence

$$F(\eta + \delta, \theta) = \int_0^\theta \int_{\mathbb{Q}_p^n} \Phi(\eta + \delta, \theta, \xi, \tau) f(\xi, \tau) d^n\xi \, d\tau$$

$$= \int_0^\theta \int_{\mathbb{Q}_p^n} \Phi(\eta + \delta, \theta, \xi + \delta, \tau) f(\xi + \delta, \tau) d^n\xi \, d\tau = F(\eta, \theta).$$

Similarly, $G(\eta + \delta, \theta) = G(\eta, \theta)$ when $\|\delta\|_p \le p^{-L}$. Thus $u_1(x, t)$, $u_2(x, t) \in \mathfrak{M}_\lambda$ uniformly in t. Thus the potentials in the expressions for $u_1(x, t)$, $u_2(x, t)$ satisfy the conditions to use the differentiation formulas given in Proposition 57. By using these formulas along with Proposition 57, (3.21)–(3.22) and (3.31), ones verifies after simple transformations that $u(x, t)$ is a solution of Cauchy problem (2.16).

Let us show that $u(x, t) \to \varphi(x)$ as $t \to 0^+$. Due to (3.28) and (3.29), it is sufficient to verify that

$$v(x, t) := \int_{\mathbb{Q}_p^n} Z(x - \xi, t; \xi, 0)\varphi(\xi)d^n\xi \to \varphi(x) \text{ as } t \to 0^+.$$

By virtue of formula (3.20), we have

$$v(x, t) = \int_{\mathbb{Q}_p^n} [Z(x - \xi, t; \xi, 0) - Z(x - \xi, t; x, 0)]\, \varphi(\xi)d^n\xi$$

$$+ \int_{\mathbb{Q}_p^n} Z(x - \xi, t; x, 0)\,[\varphi(\xi) - \varphi(x)]\, d^n\xi + \varphi(x).$$

Now, since $Z(x - \xi, t; \cdot, 0)$ and $\varphi(\cdot)$ are locally constant functions, it follows that in both integrals the integration is actually performed over the set

$$\left\{ \xi \in \mathbb{Q}_p^n; \|\xi - x\|_p \geq p^{-L} \right\}.$$

By applying (3.17) on this set, we see that both integrals tend to zero as $t \to 0^+$.

Proof of Claim C We use induction on m. The case $m = 0$ is (3.33). We assume the case m, then

$$|R_{m+1}(x, t, \xi, \tau)| \leq \int_\tau^t \int_{\mathbb{Q}_p^n} |R(x, t, \eta, \theta)|\, |R_m(\eta, \theta, \xi, \tau)|\, d^n\eta d\theta$$

$$\leq C_0(2N + 2)^{m-1} \frac{(\Gamma(v))^m}{\Gamma(mv)} \sum_{j,k=1}^{N+1} \int_\tau^t (\theta - \tau)^{(m-1)v}$$

$$\times \int_{\mathbb{Q}_p^n} \left[(\theta - \tau)^{\frac{1}{\alpha - n}} + \|\eta - \xi\|_p \right]^{-\alpha_j} \left[(t - \theta)^{\frac{1}{\alpha - n}} + \|x - \eta\|_p \right]^{-\alpha_k} d^n\eta d\theta.$$

Now by Lemma 58, with $-\sigma = (m - 1)(\alpha - n)v$, $\rho = 0$, $-n - b_2 = -\alpha_j$, $-n - b_1 = -\alpha_k$, $\beta = \alpha - n$, notice that by the condition $\alpha > n + 1$, $\beta > 1$, then we have

$$|R_{m+1}(x, t, \xi, \tau)|$$

$$\leq C_0 \left\{ (2N + 2)^{m-1} \frac{(\Gamma(v))^m}{\Gamma(mv)} \sum_{j,k=1}^{N+1} B\left(1, \frac{\alpha + (m - 1)(\alpha - n)v - \alpha_j}{\alpha - n} \right) \right.$$

$$\times \left((t-\tau)^{1/(\alpha-n)} + \|x-\xi\|_p \right)^{-\alpha_k} (t-\tau)^{\frac{(m-1)(\alpha-n)v-\alpha_j+\alpha)}{\alpha-n}}$$

$$+ B\left(\frac{\alpha-\alpha_k}{\alpha-n}, \frac{\alpha-n+(m-1)(\alpha-n)v}{\alpha-n} \right) \left((t-\tau)^{1/(\alpha-n)} + \|x-\xi\|_p \right)^{-\alpha_j}$$

$$\times (t-\tau)^{\frac{(m-1)(\alpha-n)v-\alpha_k+\alpha)}{\alpha-n}} \Bigg\},$$

where $B(\cdot,\cdot)$ denotes the Archimedean Beta function.   ∎

By using $B(z_1 + \epsilon, z_2 + \delta) \le B(z_1, z_2)$, for $\epsilon, \delta \ge 0$,

$$B\left(1, \frac{\alpha+(m-1)(\alpha-n)v-\alpha_j}{\alpha-n} \right) \le B(v, mv),$$

$$B\left(\frac{\alpha-\alpha_k}{\alpha-n}, \frac{\alpha-n+(m-1)(\alpha-n)v}{\alpha-n} \right) \le B(v, mv),$$

and

$$(t-\tau)^{\frac{(m-1)(\alpha-n)v-\alpha_r+\alpha)}{\alpha-n}} = (t-\tau)^{mv - \frac{(\alpha-n)v+\alpha_r-\alpha}{\alpha-n}} \le C(t-\tau)^{mv},$$

for $1 \le r \le N+1$, we get

$$|R_{m+1}(x,t,\xi,\tau)| \le C(2N+2)^m \frac{(\Gamma(v))^m}{\Gamma((m+1)v)} (t-\tau)^{mv}$$

$$\times \sum_{k=1}^{N+1} \left((t-\tau)^{1/(\alpha-n)} + \|x-\xi\|_p \right)^{-\alpha_k}. \qquad ∎$$

3.5   Uniqueness of the Solution

We recall that $\tilde{\mathcal{E}}$ is the $\mathbb{C}$-vector space of all functions $\varphi : \mathbb{Q}_p^n \to \mathbb{C}$, such that there exist a ball B_l^n, with l depending only on φ, and $\varphi(x+x') = \varphi(x)$ for any $x' \in B_l^n$. Notice that $\mathfrak{M}_\lambda \subset \tilde{\mathcal{E}}$ for any λ. We identify an element of $\tilde{\mathcal{E}}$ with a distribution on $\mathbb{Q}_p^n$. We now recall the following fact: $T \in \mathcal{D}'$ with $supp(T) \subset \overline{B_N^n}$ if only if $\widehat{T} \in \tilde{\mathcal{E}}$ and its parameter of local constancy is greater than $-N$, cf. [111, p. 109].

Lemma 60 $\mathbf{W}_\alpha : \tilde{\mathcal{E}} \to \tilde{\mathcal{E}}$ *is a well-defined linear operator. Furthermore,*

$$(\mathbf{W}_\alpha \varphi)(x) = -\mathcal{F}_{\xi \to x}^{-1} \left(A_{w_\alpha}(\|\xi\|_p) \mathcal{F}_{x \to \xi} \varphi \right).$$

Proof Let l be a parameter of locally constancy of φ, then

$$(\mathbf{W}_\alpha \varphi)(x) = \int_{\|y\|_p \geq p^l} \frac{\varphi(x-y) - \varphi(x)}{w_\alpha(\|y\|_p)} d^n y$$

$$= \frac{1_{\mathbb{Q}_p^n \smallsetminus B_l^n}(x)}{w_\alpha(\|x\|_p)} * \varphi(x) - \varphi(x) \left(\int_{\|y\|_p \geq p^l} \frac{d^n y}{w_\alpha(\|y\|_p)} \right).$$

Then by taking the Fourier transform in $\mathcal{D}'$:

$$\mathcal{F}(\mathbf{W}_\alpha \varphi)(\xi) = \left(\int_{\mathbb{Q}_p^n} 1_{\mathbb{Q}_p^n \smallsetminus B_l^n}(x) \frac{(\chi_p(x \cdot \xi) - 1)}{w_\alpha(\|x\|_p)} d^n x \right) (\mathcal{F}\varphi)(\xi),$$

and since $\mathcal{F}\varphi \in \mathcal{D}'$ with $supp(\mathcal{F}\varphi) \subset B_{-l}^n$,

$$\mathcal{F}(\mathbf{W}_\alpha \varphi)(\xi) = \left(\int_{\mathbb{Q}_p^n} \frac{(\chi_p(x \cdot \xi) - 1)}{w_\alpha(\|x\|_p)} d^n x \right) \mathcal{F}\varphi(\xi).$$

Therefore,

$$(\mathbf{W}_\alpha \varphi)(x) = -\mathcal{F}_{\xi \to x}^{-1} \left(A_{w_\alpha}(\|\xi\|_p) \mathcal{F}_{x \to \xi} \varphi \right) \in \tilde{\mathcal{E}}. \qquad \blacksquare$$

Take γ be a real number such that $\lambda < \gamma < \alpha_1 - n < \ldots < \alpha_N - n < \alpha - n$, and fix a integer L, and set $\psi(x) := p^{Ln} \Omega(p^L \|x\|_p) * \|x\|_p^\gamma$, then

$$\psi(x) = \begin{cases} \|x\|_p^\gamma & \text{if } \|x\|_p > p^{-L} \\ C & \text{if } \|x\|_p \leq p^{-L}, \end{cases} \tag{3.38}$$

and thus $\psi \in \tilde{\mathcal{E}}$.

Lemma 61 *With above notation, there exist positive constants C_1 and C_2 such that (i) $|(\mathbf{W}_\alpha \psi)(x)| \leq C_1 \|x\|_p^{\alpha - \gamma + n}$ and (ii) $|(\mathbf{W}_{\alpha_k} \psi)(x)| \leq C_2 \|x\|_p^{\alpha_k - \gamma + n}$, for $k = 1, \ldots, N$.*

Proof By Lemma 60,

$$(\mathbf{W}_\alpha \psi)(x) = -\mathcal{F}_{\xi \to x}^{-1} \left(A_{w_\alpha}(\|\xi\|_p) \Omega(p^{-L} \|\xi\|_p) \frac{\|\xi\|_p^{-\gamma - n}}{\Gamma_n(n + \gamma)} \right) \text{ in } \mathcal{D}',$$

where $\Gamma_n(n+\gamma) = \frac{1-p^\gamma}{1-p^{-\gamma-n}}$. Now, since $A_{w_\alpha}(\|\xi\|_p)\Omega(p^{-L}\|\xi\|_p)\frac{\|\xi\|_p^{-\gamma-n}}{\Gamma_n(n+\gamma)}$ is radial and locally integrable, by applying the formula for Fourier transform of radial function, see e.g. [111, Example 8, p. 43.],

$$
(\mathbf{W}_\alpha\varphi)(x) = \frac{-\|x\|_p^{-n}}{\Gamma_n(n+\gamma)}[(1-p^{-n})\|x\|_p^{\gamma+n}\sum_{j=0}^{\infty}A_{w_\alpha}(\|x\|_p^{-1}p^{-j})
$$

$$
\times \Omega(\|x\|_p^{-1-j}p^{-j})p^{j(\gamma+n)-jn} - A_{w_\alpha}(\|x\|_p\,p^{-j})\Omega(\|x\|_p^{-1}p^{-L+1})\|x\|_p^{\gamma+n}],
$$

as a distribution on $\mathbb{Q}_p^n \setminus \{0\}$, now by using Lemma 8 in Chap. 2

$$
|(\mathbf{W}_\alpha\varphi)(x)| \leq C'\left[(1-p^{-n})\sum_{j=0}^{\infty}p^{-j(\alpha-n)+j\gamma} - p^{(-L+1)(\alpha-n)}\right]\|x\|_p^{-\alpha+n+\gamma}.
$$

The proof of (ii) is similar.   ∎

Theorem 62 *Assume that the coefficients $a_k(x,t)$, $k = 0,1,\ldots,N$, are nonnegative bounded continuous functions, $b(x,t)$ is a bounded continuous function, $\alpha > n+1$, $0 \leq \lambda < \alpha_1 - n$ (if $a_1(x,t) = \cdots a_k(x,t) \equiv 0$, we shall suppose that $0 \leq \lambda < \alpha - n$) and $u(x,t)$ is a solution of Cauchy problem (3.15) with $f(x,t) = \varphi(x) \equiv 0$ that belongs to class $\mathfrak{M}_\lambda$. Then $u(x,t) \equiv 0$.*

Proof We may assume that $b(x,t) \geq 0$, otherwise we take $u(x,t)e^{\lambda t}$ with $\lambda > b(x,t)$. We prove that $u(x,t) \geq 0$. By contradiction, suppose that $u(x',t') < 0$, for some $x' \in \mathbb{Q}_p^n$ and $t' \in (0,T]$. By Lemma 61, it follows that $(\mathbf{W}_\alpha\psi)(x)$ and $(\mathbf{W}_{\alpha_k}\psi)(x) \to 0$ as $\|x\|_p \to \infty$, and thus

$$
M := \sup_{\substack{0 \leq t \leq T, \\ x \in \mathbb{Q}_p^n}} \left\{a_0(x,t)\,|(\mathbf{W}_\alpha\psi)(x)| + \sum_{k=1}^{N}a_k(x,t)\,|(\mathbf{W}_{\alpha_k}\psi)(x)|\right\} < \infty.
$$

We pick $\rho > 0$ such that $u(x',t') + T\rho < 0$, and then $\sigma > 0$ such that

$$
u(x',t') + T\rho + \sigma\psi(x') < 0 \tag{3.39}
$$

$$
\rho - \sigma M < 0. \tag{3.40}
$$

We now consider the function $v(x,t) := u(x,t) + t\rho + \sigma\psi(x)$. From (3.39), it follows that $v(x',t') < 0$, so that

$$
\inf_{0 \leq t \leq T, x \in \mathbb{Q}_p^n} v(x,t) < 0.
$$

Now, since $u(x, t) \in \mathfrak{M}_\lambda$, $\displaystyle\lim_{\|x\|_p \to \infty} \frac{u(x,t)}{\psi(x)} = 0$ and thus $\displaystyle\lim_{\|x\|_p \to \infty} v(x, t) > 0$ for any $t > 0$. This implies that there exist $x_0 \in \mathbb{Q}_p^n$ and $t_0 \in (0, T]$, such that

$$\inf_{0 \le t \le T, \, x \in \mathbb{Q}_p^n} v\,(x, t) = \min_{0 \le t \le T, \, x \in \mathbb{Q}_p^n} v(x, t) = v(x_0, t_0) < 0,$$

and thus, by formula (2.3), $(\mathbf{W}_\alpha v)(x_0, t_0) \ge 0$, $(\mathbf{W}_{\alpha_k} v)(x_0, t_0) \ge 0$ for all k, and $\frac{\partial v}{\partial t}(x_0, t_0) \le 0$, hence

$$\frac{\partial v}{\partial t}(x_0, t_0) - a_0(x, t)(\mathbf{W}_\alpha v)(x_0, t_0) - \sum_{k=1}^N a_k(x, t)(\mathbf{W}_{\alpha_k} v)(x_0, t_0) + b(x, t)v(x_0, t_0) < 0.$$

Now, by (3.40),

$$\frac{\partial v}{\partial t}(x, t) - a_0(x, t)(\mathbf{W}_\alpha v)(x, t) - \sum_{k=1}^N a_k(x, t)(\mathbf{W}_{\alpha_k} v)(x, t) + b(x, t)v(x, t)$$

$$= \rho - \sigma \left[a_0(x, t)(\mathbf{W}_\alpha \psi)(x) + \sum_{k=1}^N a_k(x, t)(\mathbf{W}_{\alpha_k} \psi)(x) \right] + b(x, t)\left[\rho t + \sigma \psi(x)\right]$$

$$\ge \rho - \sigma M > 0.$$

We have obtained a contradiction, thus $u(x, t) \ge 0$. Finally taking $-u(x, t)$ instead of $u(x, t)$, we conclude that $u(x, t) \equiv 0$. ∎

3.6 Markov Processes

In this section we show that the fundamental solution $\Lambda(x, t, \xi, \tau)$ of Cauchy problem (3.15) is the transition density of a Markov process. We need some preliminary results.

Lemma 63 *If the coefficients $a_k(x, t)$ and $b(x, t)$ are nonnegative, then $\Lambda(x, t, \xi, \tau) \ge 0$.*

Proof It is sufficient to show that $u(x, t) = \int_{\mathbb{Q}_p^n} \Lambda(x, t, \xi, \tau)\varphi(\xi)d^n\xi \ge 0$, where $u(x, t)$ is the solution of Cauchy problem (3.15) with $f(x, t) \equiv 0$, and initial condition $u(x, 0) = \varphi(x) \ge 0$ with $\varphi \in \mathcal{D}(\mathbb{Q}_p^n)$. From (3.28), (3.29), and Lemma 44 (iii), it follows that

$$u(x, t) \to 0 \text{ as } \|x\|_p \to \infty. \tag{3.41}$$

Now, if $u(x, t) < 0$, then there exist $x_0 \in \mathbb{Q}_p^n$ and $t_0 \in (0, T]$ such that

$$\inf_{0 \le t \le T, \, x \in \mathbb{Q}_p^n} u\,(x, t) = u(x_0, t_0) < 0. \tag{3.42}$$

This implies that $(\mathbf{W}_\alpha u)(x_0, t_0) \ge 0$, $(\mathbf{W}_{\alpha_k} u)(x_0, t_0) \ge 0$ for all k, and $\frac{\partial u}{\partial t}(x_0, t_0) \le 0$. On the other hand,

$$\frac{\partial u}{\partial t}(x, t) - a_0(x, t)(\mathbf{W}_\alpha u)(x, t) - \sum_{k=1}^{N} a_k(x, t)(\mathbf{W}_{\alpha_k} u)(x, t) = 0.$$

By using the uniform parabolicity condition $a_0(x, t) \ge \mu > 0$, we get $(\mathbf{W}_\alpha u)(x_0, t_0) = 0$, then by (2.3), $u(x, t_0)$ is constant, and by (3.41), $u(x, t_0) \equiv 0$, which contradicts (3.42). ∎

Lemma 64 *If $b(x, t) \equiv 0$, then $\int_{\mathbb{Q}_p^n} \Lambda(x, t, \xi, \tau) d^n\xi = 1$.*

Proof By integrating (3.30) in the variable ξ over whole the space $\mathbb{Q}_p^n$, and by using Lemma 50 (iii), we have $\frac{\partial}{\partial t} \left(\int_{\mathbb{Q}_p^n} \Lambda(x, t, \xi, \tau) d^n\xi \right) = 0$, thus $\int_{\mathbb{Q}_p^n} \Lambda(x, t, \xi, \tau) d^n\xi$ is independent of t. Now, by integrating (3.28) over whole space $\mathbb{Q}_p^n$ in variable ξ and by using Lemma 44 (iv), we have

$$\int_{\mathbb{Q}_p^n} \Lambda(x, t, \xi, \tau) d^n\xi = 1 + \int_\tau^t \int_{\mathbb{Q}_p^n} \int_{\mathbb{Q}_p^n} Z(x - \eta, t - \theta, \eta, \theta) \phi(\eta, \theta, \xi, \tau) d^n\eta \, d^n\xi \, d\theta.$$

The result is obtained by taking $t = \tau$ in the above formula. ∎

Lemma 65 *If $b(x, t) \equiv 0$ and $f(x, t) \equiv 0$, then the function $\Lambda(x, t, \xi, \tau)$ satisfies the following property:*

$$\Lambda(x, t, \xi, \tau) = \int_{\mathbb{Q}_p^n} \Lambda(x, t, y, \sigma) \Lambda(y, \sigma, \xi, \tau) d^n y. \tag{3.43}$$

Proof Consider the following initial value problem:

$$\begin{cases} \frac{\partial u}{\partial t}(x, t) - a_0(x, t)(\mathbf{W}_\alpha u)(x, t) - (\tilde{W}u)(x, t) = 0 \\ u\,(x, \tau) = \varphi(x), \; x \in \mathbb{Q}_p^n \text{ and } t \in (\tau, \sigma], \end{cases} \tag{3.44}$$

by Theorem 59, $u(x, \sigma) = \int_{\mathbb{Q}_p^n} \Lambda(x, \sigma, \xi, \tau)\varphi(\xi)d^n\xi$. Now consider

$$
\begin{cases}
\frac{\partial u}{\partial t}(x, t) - a_0(x, t)(\mathbf{W}_\alpha u)(x, t) - (\tilde{W}u)(x, t) = 0 \\
u(x, \sigma) = \int\limits_{\mathbb{Q}_p^n} \Lambda(x, \sigma, \xi, \tau)\varphi(\xi)d^n\xi, \ x \in \mathbb{Q}_p^n, t \in (\sigma, T], \text{ with } \tau < \sigma < T,
\end{cases}
$$

$$(3.45)$$

by Theorem 59 and Fubini's theorem, the solution of (3.45) is given by

$$
u(x, t) = \int\limits_{\mathbb{Q}_p^n} \left(\int\limits_{\mathbb{Q}_p^n} \Lambda(x, t, y, \sigma)\Lambda(y, \sigma, \xi, \tau)d^n y \right) \varphi(\xi)d^n\xi.
$$

On the other hand, (3.45) is equivalent to

$$
\begin{cases}
\frac{\partial u}{\partial t}(x, t) - a_0(x, t)(\mathbf{W}_\alpha u)(x, t) - (\tilde{W}u)(x, t) = 0 \\
u(x, \tau) = \varphi(x), \ x \in \mathbb{Q}_p^n, \ t \in (\tau, T],
\end{cases}
$$

$$(3.46)$$

which has solution given by $u(x, t) = \int_{\mathbb{Q}_p^n} \Lambda(x, t, \xi, \tau)\varphi(\xi)d^n\xi$. Now, by Theorem 62,

$$
\int\limits_{\mathbb{Q}_p^n} \Lambda(x, t, \xi, \tau)\varphi(\xi)d^n\xi = \int\limits_{\mathbb{Q}_p^n} \left(\int\limits_{\mathbb{Q}_p^n} \Lambda(x, t, y, \sigma)\Lambda(y, \sigma, \xi, \tau)d^n y \right) \varphi(\xi)d^n\xi,
$$

for any test function φ, which implies (3.43).   ∎

Theorem 66 *If the coefficients $a_k(x, t)$, $k = 1, \cdots, N$, are nonnegative bounded continuous functions, $b(x, t) \equiv 0$, $\alpha > n + 1$, $0 \le \lambda < \alpha_1 - n$ (if $a_1(x, t) = \cdots a_k(x, t) \equiv 0$, we shall suppose that $0 \le \lambda < \alpha - n$), and $f(x, t) \equiv 0$, then the fundamental solution $\Lambda(x, t, \xi, \tau)$ is the transition density of a bounded right-continuous Markov process without second kind discontinuities.*

Proof The result follows from [39, Theorem 3.6] by using Lemmas 63, 64, 65, and (3.28)–(3.29), and Lemma 44 (iii).   ∎

3.7 The Cauchy Problem Is Well-Posed

In this section, we study the continuity of the solution of Cauchy problem (3.15) with respect to $\varphi(x)$ and $f(x, t)$. We assume that the coefficients $a_k(x, t)$, $k = 0, 1, \ldots, N$ are nonnegative bounded continuous functions, $b(x, t)$ is a bounded continuous function, $0 \leq \lambda < \alpha_1 - n$ (if $a_1(x,t) = \ldots = a_k(x,t) \equiv 0$, we shall suppose that $0 \leq \lambda < \alpha - n$), $\varphi(x) \in \mathfrak{M}_\lambda$ and $f(x,t) \in \mathfrak{M}_\lambda$, uniformly in t, with $0 \leq \lambda < \alpha_1 - n$.

We identify $\mathfrak{M}_\lambda$ with the $\mathbb{R}$-vector space of all the functions "$\phi(x,t) \in \mathfrak{M}_\lambda$, uniformly in t," and introduce on $\mathfrak{M}_\lambda$ the following norm:

$$\|\phi\|_{\mathfrak{M}_\lambda} := \sup_{t \in [0,T]} \sup_{x \in \mathbb{Q}_p^n} \left| \frac{\phi(x,t)}{1 + \|x\|_p^\lambda} \right|.$$

From now on, we consider $\mathfrak{M}_\lambda$ as topological vector space with the topology induced by $\|\cdot\|_{\mathfrak{M}_\lambda}$. We also consider $\mathfrak{M}_\lambda \times \mathfrak{M}_\lambda$ as topological vector space with the topology induced by the norm $\|\cdot\|_{\mathfrak{M}_\lambda} + \|\star\|_{\mathfrak{M}_\lambda}$.

Theorem 67 *With the above hypotheses, consider the following operator:*

$$\mathfrak{M}_\lambda \times \mathfrak{M}_\lambda \xrightarrow{\boldsymbol{L}} \mathfrak{M}_\lambda$$
$$(\varphi(x), f(x,t)) \to u(x,t),$$

where $u(x,t)$ is given by (3.27). Then

$$\|u(x,t)\|_{\mathfrak{M}_\lambda} \leq C \left(\|\varphi(x)\|_{\mathfrak{M}_\lambda} + \|f(x,t)\|_{\mathfrak{M}_\lambda} \right),$$

i.e. $\boldsymbol{L}$ is a continuous operator.

Proof We write $u(x,t) = u_1(x,t) + u_2(x,t)$ where

$$u_1(x,t) = \int_0^t \int_{\mathbb{Q}_p^n} \Lambda(x,t,\xi,\tau) f(\xi,\tau) d^n\xi d\tau \text{ and}$$

$$u_2(x,t) = \int_{\mathbb{Q}_p^n} \Lambda(x,t,\xi,0)\varphi(\xi) d^n\xi,$$

as before. Now

$$|u_1(x,t)| \leq \int_0^t \int_{\mathbb{Q}_p^n} |\Lambda(x,t,\xi,\tau)| \, |f(\xi,\tau)| \, d^n\xi d\tau$$

$$\leq \|f(x,t)\|_{\mathfrak{M}_\lambda} \left\{ \int_0^t \int_{\mathbb{Q}_p^n} |\Lambda(x,t,\xi,\tau)| \, d^n\xi d\tau + \int_0^t \int_{\mathbb{Q}_p^n} |\Lambda(x,t,\xi,\tau)| \, \|\xi\|_p^\lambda \, d^n\xi d\tau \right\},$$

by (3.28)–(3.29), (3.17) and Proposition 45,

$$
|u_1(x,t)| \leq C_0 \, \|f(x,t)\|_{\mathfrak{M}_\lambda} \left\{ \int_0^t (t-\tau)^{1+\frac{n-\alpha}{\alpha-n}} \, d\tau + \int_0^t (t-\tau)^{2-\frac{\lambda}{\alpha-n}+\frac{n-\alpha}{\alpha-n}} \, d\tau \right.
$$

$$
+ \sum_{k=1}^{N+1} \int_0^t (t-\tau)^{1+\frac{n-\alpha_k}{\alpha-n}} \, d\tau + \left(1 + \|x\|_p^\lambda\right) \int_0^t (t-\tau)^{1+\frac{n-\alpha}{\alpha-n}} \, d\tau
$$

$$
+ \left(1 + \|x\|_p^\lambda\right) \int_0^t (t-\tau)^{2-\frac{\lambda}{\alpha-n}+\frac{n-\alpha}{\alpha-n}} \, d\tau
$$

$$
\left. + \left(1 + \|x\|_p^\lambda\right) \sum_{k=1}^{N+1} \int_0^t (t-\tau)^{1+\frac{n-\alpha_k}{\alpha-n}} \, d\tau \right\}
$$

$$
\leq \|f(x,t)\|_{\mathfrak{M}_\lambda} \left\{ C_1(T) + C_2(T) \left(1 + \|x\|_p^\lambda\right) \right\}.
$$

Hence,

$$
\left| \frac{u_1(x,t)}{1 + \|x\|_p^\lambda} \right| \leq \|f(x,t)\|_{\mathfrak{M}_\lambda} \left\{ \frac{C_1(T)}{1 + \|x\|_p^\lambda} + C_2(T) \right\}.
$$

In the same form, one shows that

$$
\left| \frac{u_2(x,t)}{1 + \|x\|_p^\lambda} \right| \leq \|\varphi(x)\|_{\mathfrak{M}_\lambda} \left\{ \frac{C_1'(T)}{1 + \|x\|_p^\lambda} + C_2'(T) \right\},
$$

therefore $\|u(x,t)\|_{\mathfrak{M}_\lambda} \leq C \left(\|\varphi(x,t)\|_{\mathfrak{M}_\lambda} + \|f(x,t)\|_{\mathfrak{M}_\lambda} \right).$ ∎

Chapter 4
Parabolic-Type Equations and Markov Processes on Adeles

4.1 Introduction

In this chapter we study the Cauchy problem for parabolic type pseudodifferential equations over the rings of finite adeles and adeles involving a generalization of the Taibleson operator [5, 98]. For the sake of simplicity we formulated all our results for finite adeles and adeles on $\mathbb{Q}$, however all the results are still valid if the field of rational numbers $\mathbb{Q}$ is replaced by a global field, i.e. by an algebraic number field, or by the function field of an algebraic curve over a finite field. The results presented in this chapter were obtained by Torba and Zúñiga-Galindo in [106].

There are several motivations for this line of research. As we already pointed out, during the last 30 years the interest on stochastic processes on p-adics and adeles, see e.g. [1, 8–14, 19, 20], [36, and references therein], [53, 62], [80, and references therein], [90, 98, 108], [111, and references therein], [116, 122], among others. Another two motivations for studying pseudodifferential equations on adeles are the following. In [51] Haran established a connection between explicit formulas for the Riemann zeta function and adelic pseudodifferential operators, see also [29]. In [85] Manin posed the conjecture that the physical space is adelic, which can be considered as an extension of the Volovich conjecture on the non Archimedean nature of the physical space at the Planck scale [109, 112, 113]. This conjecture conducts naturally to consider models involving partial differential equations on adelic spaces. Some preliminary results such as studying pseudodifferential operators (in an adelic setting) are presented in [35, and references therein], [37, 51, 72, 74, 91].

The chapter is organized as follows. In Sect. 4.2 we summarize some well-known results on adelic analysis. In Sects. 4.3 and 4.4 we introduce metric structures on the rings of finite adeles and adeles, see Propositions 71, 95. These metric structures induce the adelic topology and are naturally connected with the Fourier transform. In addition, they allow us to use classical results on Markov processes, see e.g. [39]. We compute the Fourier transform of radial functions defined on the ring of

© Springer International Publishing AG 2016
W.A. Zúñiga-Galindo, *Pseudodifferential Equations Over Non-Archimedean Spaces*, Lecture Notes in Mathematics 2174,
DOI 10.1007/978-3-319-46738-2_4

finite adeles, see Theorem 79, and we introduce adelic analogues of the Taibleson operators and Lizorkin spaces of the second kind and prove some basic properties of them. In Sect. 4.5 we study the heat kernels on the ring of finite adeles, see Definition 101 and Theorem 105. We give an 'explicit formula' for the heat kernel as a series involving Chebyshev type functions, i.e. products of powers of primes, some arithmetic operators and exponential functions depending on t, see Proposition 102. We require the prime number theorem to establish the existence of the adelic heat kernels, see Proposition 100. In Sect. 4.6 we show that the adelic heat kernels are the transition functions of Markov processes, see Theorem 108. In Sects. 4.8 and 4.9 we study the heat kernels on the ring of adeles, see Definition 119 and Theorem 120, and show that the heat kernels are the transition functions of Markov processes, see Theorem 123. In Sects. 4.7 and 4.10 we study Cauchy problems for parabolic type equations involving adelic versions of the Taibleson operator. We show that these problems are well-posed and find explicit formulas for the solutions of homogeneous and non-homogeneous equations, see Proposition 112, Theorems 114, 117, 118, Proposition 126 and Theorems 127, 128.

4.2 Preliminaries

In this section we fix the notation and collect some basic results on p-adic and adelic analysis that we will use through this chapter. For a detailed exposition on p-adic and adelic analysis the reader may consult [5, 50, 80, 93, 105, 111].

4.2.1 Adeles on $\mathbb{Q}$

Along this chapter, the variables p, q will denote 'primes', including 'the infinite prime', denoted by ∞. To each prime p corresponds an absolute value $|\cdot|_p$ on $\mathbb{Q}$, with $|\cdot|_\infty$ corresponding to the usual Euclidean norm. In addition, $\mathbb{Q}_p$ denotes the completion of $\mathbb{Q}$ with respect to $|\cdot|_p$, note that $\mathbb{Q}_\infty = \mathbb{R}$. If p is a finite prime, we denote by dx_p the Haar measure of the topological group $(\mathbb{Q}_p, +)$ normalized by the condition $\mathrm{vol}(\mathbb{Z}_p) = 1$. Along this chapter, for p finite, we denote the p-adic order of $x \in \mathbb{Q}_p$ by $ord_p(x)$ instead of $ord(x)$, since p is varying over all the primes.

The *ring of adeles over* $\mathbb{Q}$, denoted $\mathbb{A}$, is defined by

$$\mathbb{A} = \big\{ (x_\infty, x_2, x_3, \ldots) \, ; x_p \in \mathbb{Q}_p, \text{ and } x_p \in \mathbb{Z}_p \text{ for all but finitely many } p \big\}.$$

Alternatively, we can define $\mathbb{A}$ as *the restricted product* of the $\mathbb{Q}_p$ with respect to the $\mathbb{Z}_p$. The componentwise addition and multiplication give to $\mathbb{A}$ a ring structure. Furthermore, $\mathbb{A}$ can be made into a locally compact topological ring by taking as a base for the topology, certainly the *restricted product topology*, all the sets of the form $U \times \prod_{p \notin S} \mathbb{Z}_p$ where S is any finite set of primes containing ∞, and U is any open subset in $\prod_{p \in S} \mathbb{Q}_p$.

The restricted product topology is not equal to the product topology. However, the following relation holds. Take S as before and consider the group

$$G_S = \prod_{p \in S} \mathbb{Q}_p \times \prod_{p \notin S} \mathbb{Z}_p.$$

Then, the product topology on G_S is identical to the one induced by the restricted product topology on G_S, thus G_S is a locally compact subgroup of $\mathbb{A}$, and the locally compact topological group $(\mathbb{A}, +)$ has a Haar measure, denoted $dx_\mathbb{A}$, which coincides on G_S with the product measure $\prod_p dx_p$, where dx_∞ is the Lebesgue measure of $\mathbb{R}$. We also note that any set of the form

$$\prod_{p \in S} p^{l_p} \mathbb{Z}_p \times \prod_{p \notin S} \mathbb{Z}_p, \tag{4.1}$$

where l_p are arbitrary integers, is a compact subset of $\mathbb{A}$.

The *ring of finite adeles* over $\mathbb{Q}$, denoted $\mathbb{A}_f$, is defined by

$$\mathbb{A}_f = \left\{ (x_2, x_3, \ldots) ; x_p \in \mathbb{Q}_p, \text{ and } x_p \in \mathbb{Z}_p \text{ for all but finitely many } p \right\}.$$

From now on, we consider $\mathbb{A}_f$ as a topological ring with respect to the restricted product topology. Then $\mathbb{A} = \mathbb{R} \times \mathbb{A}_f$. Since $(\mathbb{A}_f, +)$ is a locally compact topological group, it has a Haar measure, denoted $dx_{\mathbb{A}_f}$, which agrees with the product measure $\prod_{p < \infty} dx_p$ on open subgroups of type

$$\prod_{p \leq N} \mathbb{Q}_p \times \prod_{p > N} \mathbb{Z}_p, \quad \text{with } N \in \mathbb{N}.$$

Furthermore $dx_\mathbb{A} = dx_\infty dx_{\mathbb{A}_f}$. For a detailed presentation of the integration theory on $\mathbb{A}$ and $\mathbb{A}_f$ see [50, Chapter 1], see also [93, 115].

Remark 68 The multiplicative group of $\mathbb{A}$, denoted $\mathbb{A}^\times$, is called the group of ideles over $\mathbb{Q}$, it is defined as

$$\mathbb{A}^\times = \left\{ (x_\infty, x_2, x_3, \ldots) ; x_p \in \mathbb{Q}_p^\times, \text{ and } x_p \in \mathbb{Z}_p^\times \text{ for all but finitely many } p \right\},$$

where $\mathbb{Z}_p^\times = \left\{ u \in \mathbb{Z}_p ; |u|_p = 1 \right\}$ denotes the groups of units of $\mathbb{Z}_p$. The group of finite ideles over $\mathbb{Q}$, denoted $\mathbb{A}_f^\times$, is defined as

$$\mathbb{A}_f^\times = \left\{ (x_2, x_3, \ldots) ; x_p \in \mathbb{Q}_p^\times, \text{ and } x_p \in \mathbb{Z}_p^\times \text{ for all but finitely many } p \right\}.$$

The ideles over $\mathbb{Q}$ form a locally compact topological group but its topology is not induced by the topology of $\mathbb{A}$.

In this chapter we work exclusively with complex valued functions on adeles. Having complex valued functions defined on a locally compact topological group, we have the notion of continuous function and may use the functional spaces $L^\varrho(\mathbb{A}_f)$ and $L^\varrho(\mathbb{A})$, $\varrho \geq 1$ defined in the standard way.

For studying solutions of parabolic equations we need notations for several spaces of functions which depend on time and adelic (space) variables. We denote by:

(i) $C(I, X)$ the space of continuous functions u on a time interval I with values in X;

(ii) $C^1(I, X)$ the space of continuously differentiable functions u on a time interval I such that $u' \in C(I, X)$;

(iii) $L^1(I, X)$ the space of measurable functions u on I with values in X such that $\|u\|$ is integrable;

(iv) $W^{1,1}(I, X)$ the space of measurable functions u on I with values in X such that $u' \in L^1(I, X)$.

4.2.2 Fourier Transform on Adeles

Let p be a finite prime, we denote by $\chi_p : \mathbb{Q}_p \to \mathbb{C}^\times$ the additive character defined by

$$\chi_p(x_p) = \exp(2\pi i \{x_p\}),$$

where

$$\{x_p\} := \begin{cases} \sum_{i=-k}^{-1} a_i p^i & \text{if } x_p = \sum_{i=-k}^{\infty} x_{p,i} p^i \text{ with } k > 0 \text{ and } 0 \leq x_{p,i} \leq p-1 \\ 0 & \text{otherwise}, \end{cases}$$

as before. We recall that a function $f_p : \mathbb{Q}_p \to \mathbb{C}$ which is locally constant with compact support is called a *Bruhat-Schwartz function*. The space of such functions is denoted as $\mathcal{D}(\mathbb{Q}_p)$, as before.

For $f_p \in \mathcal{D}(\mathbb{Q}_p)$, its Fourier transform $\widehat{f}_p$ is defined by

$$\widehat{f}_p(\xi_p) = \int_{\mathbb{Q}_p} \chi_p(x_p \xi_p) f_p(x_p) \, dx_p.$$

The Fourier transform induces a linear isomorphism of $\mathcal{D}(\mathbb{Q}_p)$ onto $\mathcal{D}(\mathbb{Q}_p)$ satisfying $\widehat{\widehat{f}}_p(\xi_p) = f_p(-\xi_p)$.

In the case $p = \infty$ the additive character is defined by $\chi_\infty (x_\infty) := \exp(2\pi i x_\infty)$. Let $\mathcal{D}(\mathbb{R})$ denote the Schwartz space. The Fourier transform of $f_\infty \in \mathcal{D}(\mathbb{R})$, denoted $\widehat{f_\infty}$, is defined by

$$\widehat{f_\infty} (\xi_\infty) = \int_{\mathbb{R}} \chi_\infty (-x_\infty \xi_\infty) f_\infty (x_\infty)\, dx_\infty.$$

The Fourier transform induces a linear isomorphism of $\mathcal{D}(\mathbb{R})$ onto $\mathcal{D}(\mathbb{R})$. By choosing a suitable multiple of the Lebesgue measure, the Fourier transform satisfies $\widehat{\widehat{f_\infty}} (\xi_\infty) = f_\infty (-\xi_\infty)$.

The *additive adelic character* $\chi : \mathbb{A} \to \mathbb{C}$ is defined by

$$\chi (x) = \prod_p \chi_p (x_p) \qquad \text{for } x = (x_\infty, x_2, x_3, \ldots).$$

Remark 69 By abuse of notation we will denote also by χ, the additive character on $\mathbb{A}_f$, defined by

$$\chi (x) = \prod_{p < \infty} \chi_p (x_p) \qquad \text{for } x = (x_2, x_3, \ldots).$$

An adelic function is said to be *Bruhat-Schwartz* if it can be expressed as a finite linear combination, with complex coefficients, of factorizable functions $f = \prod_{p \leq \infty} f_p$, where f_p satisfies the following conditions: (A1) $f_\infty \in \mathcal{D}(\mathbb{R})$; (A2) $f_p \in \mathcal{D}(\mathbb{Q}_p)$ for $p < \infty$; (A3) f_p is the characteristic function of $\mathbb{Z}_p$ for all but finitely many $p < \infty$. The adelic space of Bruhat-Schwartz functions is denoted as $\mathcal{D}(\mathbb{A})$. The space of Bruhat-Schwartz functions $\mathcal{D}(\mathbb{A}_f)$ is defined in a similar form except that only conditions A2 and A3 are required.

The Fourier transform of a factorizable adelic Bruhat-Schwartz function is defined by

$$\hat{f}(\xi) = \int_{\mathbb{A}} \chi(x\xi) f(x)\, dx_{\mathbb{A}}$$
$$= \int_{\mathbb{Q}_\infty} f_p (x_\infty) \chi_\infty (-x_\infty \xi_\infty)\, dx_\infty \prod_{p<\infty} \int_{\mathbb{Q}_p} f_p (x_p) \chi_p (x_p \xi_p)\, dx_p. \qquad (4.2)$$

This definition may be extended to arbitrary adelic Bruhat-Schwartz functions by linearity. The Fourier transform gives a linear isomorphism of $\mathcal{D}(\mathbb{A})$ to $\mathcal{D}(\mathbb{A})$ satisfying $\hat{\hat{f}}(\xi) = f(-\xi)$. Analogous definitions and results hold for the Fourier transform on $\mathbb{A}_f$. The Fourier transform may be extended to the space $L^2(\mathbb{A})$ (or to $L^2(\mathbb{A}_f)$), where it is a unitary operator and the Steklov–Parseval equality holds.

We will also use the notation $\mathcal{F}\varphi$ for the Fourier transform and $\mathcal{F}^{-1}\varphi$ for the inverse Fourier transform. We used as a main reference for this section [50, Chapter 1], see also [72, 93, 115]. We notice that our definition for the adelic Fourier transform is equivalent to the one given in [50, Chapter 1].

Since $\mathbb{A}$ (resp. $\mathbb{A}_f$) is a locally compact topological group, a convolution operation between functions is also defined on $\mathcal{D}(\mathbb{A})$ and $L^2(\mathbb{A})$ (resp. $\mathcal{D}(\mathbb{A}_f)$ and $L^2(\mathbb{A}_f)$). It is connected with the Fourier transform in the usual way, see e.g. [100] for details.

4.3 Metric Structures, Distributions and Pseudodifferential Operators on $\mathbb{A}_f$

4.3.1 A Structure of Complete Metric Space for the Finite Adeles

In the previous section the restricted product topology on adeles was described. In this section, we show that for the finite adeles the topology is metrizable and present a non-Archimedean metric on $\mathbb{A}_f$. Moreover, in this metric each ball is a compact set and the Fourier transform of a radial function is again a radial function. Hence, despite of the complicated form of the presented metric we believe that it is natural for the ring of finite adeles.

Consider the following two functions:

$$\|x\|_1 := \max_p |x_p|_p, \qquad x \in \mathbb{A}_f, \tag{4.3}$$

and

$$\|x\|_0 := \max_p \frac{|x_p|_p}{p}, \qquad x \in \mathbb{A}_f. \tag{4.4}$$

Both functions are well defined and may be used to introduce a metric on $\mathbb{A}_f$. However, the topology induced by the metric $\|x - y\|_1$ does not coincide with the restricted product topology which may be easily seen from the following example. The sequence of adeles $x^{(k)} := (\underbrace{0, 0, \ldots, 0}_{k-1}, 1, 0, \ldots)$, $k \in \mathbb{N}$ converges to 0 in the restricted product topology, but does not converge in the metric generated by $\|\cdot\|_1$. The metric $\|x - y\|_0$ induces the same topology as the restricted product topology, however it does not satisfy the above mentioned properties. For instance, with respect to this metric only balls of radiuses less than 1 are compact, and the Fourier transform of a radial function is not necessary a radial function. We left checking of these statements to reader, all required proofs may be obtained similarly to the proofs in this chapter.

To overcome the mentioned problems we define the function:

$$\|x\| = \begin{cases} \|x\|_0 & \text{if } x \in \prod_p \mathbb{Z}_p \\ \|x\|_1 & \text{if } x \notin \prod_p \mathbb{Z}_p, \end{cases} \tag{4.5}$$

for arbitrary $x \in \mathbb{A}_f$. Note that $\|x\|_0 \leq \|x\| \leq \|x\|_1$ for any $x \in \mathbb{A}_f$. We now introduce the function (our metric)

$$\rho(x, y) := \|x - y\|, \qquad x, y \in \mathbb{A}_f. \tag{4.6}$$

Remark 70

(i) The range of values of the function ρ coincides with the set $\{0\} \cup \{p^j; p \text{ is prime}, j \in \mathbb{Z} \setminus \{0\}\}$.
(ii) Later on we will use the following function

$$[[t]] := \begin{cases} [t] & \text{if } t \geq 0 \\ [t] + 1 & \text{if } t < 0, \end{cases} \tag{4.7}$$

where $[\cdot]$ denotes the integer part function, which is piecewise constant and left-continuous.

Proposition 71 *The restricted product topology on $\mathbb{A}_f$ is metrizable, the metric is given by (4.6). Furthermore, $(\mathbb{A}_f, \rho)$ is a complete non-Archimedean metric space.*

Proof The fact that $\rho(x, y)$ is a non-Archimedean metric is a consequence of the fact that

$$\|x + y\| \leq \max\{\|x\|, \|y\|\}, \qquad x, y \in \mathbb{A}_f,$$

which can be checked easily case by case. For any $x, y \in \mathbb{A}_f$ we have $\|x - y\|_0 \leq \rho(x, y) \leq \|x - y\|_1$, hence we may reduce the number cases to check that ρ is a non-Archimedean distance to the only case, when $x - y \notin \prod_p \mathbb{Z}_p$, $x - z \in \prod_p \mathbb{Z}_p$ and $y - z \notin \prod_p \mathbb{Z}_p$. In this case the proof follows from the equality

$$\max\{\rho(x, z), \rho(z, y)\} = \max\{\|x - z\|_0, \|z - y\|_1\} = \|z - y\|_1$$
$$= \max\{\|x - z\|_1, \|z - y\|_1\}.$$

We now show that $(\mathbb{A}_f, \rho)$ is a complete metric space. Let $x^{(n)} = \left(x_p^{(n)}\right)_p$ be a Cauchy sequence in $\mathbb{A}_f$ with respect to ρ. Since we have coordinate-wise convergence, we may define $\tilde{x}_p := \lim_{n \to \infty} x_p^{(n)}$ in $\mathbb{Q}_p$ and $\tilde{x} := (\tilde{x}_p)_p$. We assert that $\tilde{x} \in \mathbb{A}_f$. Indeed, $\rho(x^{(n)}, x^{(m)}) < 1$ for all $n, m \geq M_0$, hence $\rho(x^{(n)}, x^{(m)}) = \|x^{(n)} - x^{(m)}\|_0$. Due to the properties of p-adic absolute value it follows from $\frac{|x_p - y_p|_p}{p} < 1$ that $|x_p - y_p|_p \leq 1$.

Therefore $\left|x_p^{(n_0)} - x_p^{(m)}\right|_p \leq 1$ for all p and $n_0, m \geq M_0$. Then $\left|x_p^{(n_0)} - \tilde{x}_p\right|_p \leq 1$ for all p and $n_0 \geq M_0$. Since $\left(x_p^{(n_0)}\right)_p \in \mathbb{A}_f$, there exist a constant N such that $x_p^{(n_0)} \in \mathbb{Z}_p$ for $p \geq N$. Then also $\tilde{x}_p \in \mathbb{Z}_p$ for $p \geq N$. To show that $\lim_{n \to \infty} \rho(x^{(n)}, \tilde{x}) = 0$, consider arbitrary $\epsilon > 0$ and take an integer $N' \geq N$ such that $1/N' < \epsilon$. Since $\left|x_p^{(n)} - \tilde{x}_p\right|_p \leq 1$ for all p and $n \geq M_0$, and $x_p^{(n)} \to \tilde{x}_p$ for any p, we have for n big enough

$$
\rho(x^{(n)}, \tilde{x}) = \max \left\{ \max_{p < N'} \frac{\left|x_p^{(n)} - \tilde{x}_p\right|_p}{p}, \max_{p \geq N'} \frac{\left|x_p^{(n)} - \tilde{x}_p\right|_p}{p} \right\}
$$

$$
\leq \max \left\{ \max_{p < N'} \frac{\left|x_p^{(n)} - \tilde{x}_p\right|_p}{p}, \frac{1}{N'} \right\} \leq \max \left\{ \max_{p < N'} \frac{\left|x_p^{(n)} - \tilde{x}_p\right|_p}{p}, \epsilon \right\} = \epsilon.
$$

Let $\tau_{\mathbb{A}_f}$ denote the restricted product topology on $\mathbb{A}_f$, and let τ_ρ denote the topology induced by ρ on $\mathbb{A}_f$. We want to show that $\tau_{\mathbb{A}_f} = \tau_\rho$. Set $U := \prod_p \mathbb{Z}_p$. Then the family

$$
x + yU, \quad x \in \mathbb{A}_f, y \in \mathbb{A}_f^\times,
$$

is a base for $\tau_{\mathbb{A}_f}$. We notice that in the verification of this assertion one only needs that $y \in \mathbb{Q}^\times$, and also that U coincides with $\left\{x \in \mathbb{A}_f; \rho(0, x) \leq \frac{1}{2}\right\}$ which is open in τ_ρ. If we show that $(\mathbb{A}_f, +, \cdot)$ is a topological ring with respect to τ_ρ then $x + yU \in \tau_\rho$ for any $x \in \mathbb{A}_f$, $y \in \mathbb{A}_f^\times$, i.e. $\tau_{\mathbb{A}_f} \subset \tau_\rho$. To verify that $(\mathbb{A}_f, +, \cdot)$ is a topological ring with respect to τ_ρ it is sufficient to check the continuity of the ring operations which can be established using sequences. The continuity of the addition is straightforward, for the continuity of the multiplication consider two sequences $x^{(n)} \xrightarrow{\rho} x$ with $x^{(n)} = \left(x_p^{(n)}\right)$ and $x = \left(x_p\right) \in \mathbb{A}_f$ and $y^{(n)} \xrightarrow{\rho} y$ with $y^{(n)} = \left(y_p^{(n)}\right)$ and $y = \left(y_p\right) \in \mathbb{A}_f$. We have

$$
\rho\left(x^{(n)} y^{(n)}, xy\right) = \left\| \left(x^{(n)} - x\right)\left(y^{(n)} - y\right) + x\left(y^{(n)} - y\right) + y\left(x^{(n)} - x\right) \right\|
$$

$$
\leq \max \left\{ \left\| \left(x^{(n)} - x\right)\left(y^{(n)} - y\right) \right\|, \left\| x\left(y^{(n)} - y\right) \right\|, \left\| y\left(x^{(n)} - x\right) \right\| \right\}.
$$

Since $x^{(n)} \xrightarrow{\rho} x$ then $\rho\left(x^{(n)}, x\right) < 1$ for n big enough and $x^{(n)} - x \in \prod_p \mathbb{Z}_p$, similarly $y^{(n)} - y \in \prod_p \mathbb{Z}_p$. Therefore $\left(x^{(n)} - x\right)\left(y^{(n)} - y\right) \in \prod_p \mathbb{Z}_p$ and

$$
\left\| \left(x^{(n)} - x\right)\left(y^{(n)} - y\right) \right\| = \max_p \frac{\left|x_p^{(n)} - x_p\right|_p \left|y_p^{(n)} - y_p\right|_p}{p}
$$

$$
\leq \max_p \frac{\left|y_p^{(n)} - y_p\right|_p}{p} = \rho\left(y^{(n)}, y\right) \to 0 \quad \text{as } n \to \infty.
$$

On the other hand, for $n = n(x)$ big enough $x_p\left(y_p^{(n)} - y_p\right) \in \mathbb{Z}_p$ for all p because at most a finite number of components of the adele x does not belong to $\mathbb{Z}_p$ and $y_p^{(n)} \to y_p$ for every p. Hence

$$\left\| x\left(y^{(n)} - y\right) \right\| = \max_p \frac{|x_p(y_p^{(n)} - y_p)|_p}{p} \leq \left(\max_p |x_p|_p\right) \max_p \frac{|y_p^{(n)} - y_p|_p}{p}$$

$$= \left(\max_p |x_p|_p\right) \left\| y^{(n)} - y \right\| = \|x\|_1 \, \rho\left(y^{(n)}, y\right) \to 0 \quad \text{for } n \to \infty.$$

Similarly $\left\| y(x^{(n)} - x) \right\| \to 0$ as $n \to \infty$ which finishes the proof of the continuity of the multiplication.

We now show that $\tau_\rho \subset \tau_{\mathbb{A}_f}$. The family of balls

$$B_\epsilon\left(x^{(0)}\right) = \left\{ x \in \mathbb{A}_f; \rho\left(x^{(0)}, x\right) \leq \epsilon \right\}, \quad x^{(0)} = \left(x_p^{(0)}\right)_p \in \mathbb{A}_f \tag{4.8}$$

is a base for τ_ρ. We have

$$B_\epsilon\left(x^{(0)}\right) = \prod_p \left(x_p^{(0)} + p^{-\alpha_p(\epsilon)} \mathbb{Z}_p\right), \tag{4.9}$$

where $\alpha_p(\epsilon) = [[\log_p \epsilon]]$, here the function $[[\cdot]]$ is defined by 4.7. Note that for p big enough $\alpha_p(\epsilon) = 0$ and $x_p^{(0)} \in \mathbb{Z}_p$. Therefore by (4.9) we have $B_\epsilon(x^{(0)}) \in \tau_{\mathbb{A}_f}$. ∎

Corollary 72 $B_\epsilon\left(x^{(0)}\right)$ *is a compact subset for any $\epsilon > 0$.*

Proof By (4.9), $B_\epsilon(x^{(0)})$ is a translation of a compact subset $\prod_p p^{-\alpha_p(\epsilon)} \mathbb{Z}_p$, cf. (4.1). ∎

Remark 73 The following properties of the space $(\mathbb{A}_f, \rho)$ hold.

(i) $(\mathbb{A}_f, \rho)$ is σ-compact space. Indeed, consider

$$K_N := \prod_{p \leq N} p^{-N} \mathbb{Z}_p \times \prod_{p > N} \mathbb{Z}_p \quad \text{for } N \subset \mathbb{N}.$$

Then K_N is a compact subgroup with respect to $\tau_{\mathbb{A}_f}$, see e.g. [93, Section 5.1] and $\mathbb{A}_f = \cup_N K_N$.

(ii) $(\mathbb{A}_f, \rho)$ is second-countable topological space. Indeed, by applying twice the Weak Approximation Theorem, see e.g. [50, Theorem 1.4.4], one gets that $\beta + \alpha \prod_p \mathbb{Z}_p, \beta \in \mathbb{Q}, \alpha \in \mathbb{Q} \setminus \{0\}$ is a countable base for the topology of $\mathbb{A}_f$.

(iii) $(\mathbb{A}_f, \rho)$ is a semi-compact space, i.e. a locally compact Hausdorff space with a countable base.

Metric ρ allows us to introduce an adelic ball (given by (4.8)) and an adelic sphere, given by

$$S_r\left(x^{(0)}\right) = \left\{x \in \mathbb{A}_f; \rho\left(x^{(0)}, x\right) = r\right\}, \quad x^{(0)} = \left(x_p^{(0)}\right)_p \in \mathbb{A}_f. \tag{4.10}$$

Note that by Remark 70 the radius r of the adelic sphere may possess only values equal to any non-zero integer power of prime number. We now introduce some notations and compute volumes of adelic balls and adelic spheres.

Given a positive real number x, we define

$$\Phi(x) = \prod_p p^{[[\log_p x]]}, \tag{4.11}$$

where $[[\cdot]]$ is defined by (4.7), i.e. for $x \geq 1$ we take a product over all prime numbers each taken in the largest power α_p such that $p^{\alpha_p} \leq x$ and for $x < 1$ we take a product over all prime numbers each taken in the largest power α_p such that $p^{\alpha_p} \leq px$, see also (4.9). Note that only a finite number of terms in this product differs from 1 and that the function $\Phi(x)$ is non-decreasing, right-continuous and piecewise constant. Then $\Phi(x) = 1$ if $1/2 \leq x < 2$. If $x \geq 2$, $\Phi(x)$ coincides with the exponential of the second Chebyshev function $\psi(x) = \sum_p [\log_p x] \ln p = \sum_{p^k \leq x} \ln p$, where the last sum is taken over all powers of prime numbers not exceeding x.

Definition 74 For $n \in \mathbb{R}$, $n > 0$ we define the *next* and *previous non-zero power of a prime operators* as

$$n_+ = \min \left\{p^\beta; n < p^\beta, \ p \text{ prime}, \ \beta \in \mathbb{Z} \setminus \{0\}\right\}, \tag{4.12}$$

$$n_- = \max \left\{p^\beta; p^\beta < n, \ p \text{ prime}, \ \beta \in \mathbb{Z} \setminus \{0\}\right\}. \tag{4.13}$$

By using the operators $(\cdot)_-$ and $(\cdot)_+$ we can completely order the set of non-zero powers of primes. This total order will be very relevant in the next sections. To simplify notations we will write p_-^j instead of $(p^j)_-$ and p_+^j instead of $(p^j)_+$.

Lemma 75 *With the above notation, the following formulas hold:*

(i) for any prime number p and any $j \in \mathbb{Z} \setminus \{0\}$,

$$\Phi(p^{-j}) = \frac{p}{\Phi(p^j)}; \tag{4.14}$$

(ii) for any number $n = p^j$, where p is a prime and $j \in \mathbb{Z} \setminus \{0\}$, we have

$$(n_-)_+ = n, \ (n_+)^{-1} = (n^{-1})_-, \ (n_+)_- = n, \ (n_-)^{-1} = (n^{-1})_+;$$

(iii) *for any prime number p and any $j \in \mathbb{Z} \setminus \{0\}$,*

$$\Phi\left(p^j_-\right) = \frac{\Phi(p^j)}{p}. \tag{4.15}$$

Proof

(i) We have to show that $\Phi(p^{-j})\Phi(p^j) = p$, thus we may assume that $j > 0$. If $p \neq q$, then $\log_q p^j > 0$ and $\{\log_q p^j\} \neq 0$, where $\{\cdot\}$ denotes the fractional part function. Hence $[[\log_q p^j]] = [\log_q p^j]$, and since $[\log_q p^{-j}] = -([\log_q p^j] + \{\log_q p^j\})$, we have $[[\log_q p^{-j}]] = -[\log_q p^j]$. Now

$$\Phi(p^{-j})\Phi(p^j) = \prod_q q^{[[\log_q p^j]]} \prod_q q^{[[\log_q p^{-j}]]}$$

$$= p^{[[\log_p p^j]]} p^{[[\log_p p^{-j}]]} \prod_{q \neq p} q^{[[\log_q p^j]]} \prod_{q \neq p} q^{[[\log_q p^{-j}]]}$$

$$= p \prod_{q \neq p} q^{[[\log_q p^j]] + [[\log_q p^{-j}]]} = p \prod_{q \neq p} q^{[\log_q p^j] - [\log_q p^j]} = p.$$

(ii) The formulas follow from definitions (4.12)–(4.13). We notice that the formulas are not valid if $n \neq p^j$. For instance, if we take $n = \sqrt{5}$, then $n_- = 3$, $n_+ = 5$, and $(n_-)_+ = 5 \neq \sqrt{5}$.

(iii) In the case $j \geq 0$, the announced formula follows from:

Claim A

$$[\log_q p^j_-] = \begin{cases} j - 1 & \text{if } q = p \\ [\log_q p^j] & \text{if } q \neq p. \end{cases}$$

Indeed,

$$\Phi(p^j) = \prod_q q^{[[\log_q p^j]]} = \prod_q q^{[\log_q p^j]} = p^j \prod_{q \neq p} q^{[\log_q p^j]}$$

$$= p p^{j-1} \prod_{q \neq p} q^{[\log_q p^j_-]} = p \prod_q q^{[\log_q p^j_-]} = p\Phi(p^j_-).$$

Claim A is verified as follows. We first notice that $p^{j-1} \leq p^j_- < p^j$ implies that $j - 1 \leq [\log_q p^j_-] < j$, now by using the left continuity of the function $[\cdot]$ we get the first case of the formula. We now set $p^j_- = q^\beta$ with q a prime number different from p. From $q^\beta < p^j < q^{\beta+1}$, one gets $\beta = [\log_q p^j]$, which establish the second case in the formula for $q \neq p$. We now consider a prime k satisfying $k \neq q \neq p$.

Take $\alpha = \left[\log_k p_-^j\right]$ and $\gamma = \left[\log_k p^j\right]$. If $\alpha < \gamma$, then $k^\alpha < q^\beta < k^\gamma < p^j$, which contradicts the fact that $p_-^j = q^\beta$. Since $\alpha \leq \gamma$, we conclude that $\alpha = \gamma$.

Finally, we show the formula in the case p^{-j}, with $j > 0$. We first notice that if $p_1^{\alpha_1}, \cdots, p_k^{\alpha_k}, p^j$ are all the prime powers $\leq p^j$, i.e. if $\Phi(p^j) = p_1^{\alpha_1} \ldots p_k^{\alpha_k} p^j$, then

$$
p_+^j = \begin{cases} p_t^{1+\alpha_t} & \text{for some } t \in \{1, \cdots, k\} \\[2mm] q & \text{for some prime } q \neq p_l,\, l \in \{1, \cdots, k\}, \end{cases}
$$

and

$$
\Phi(p_+^j) = \begin{cases} p_1^{\alpha_1} \ldots p_t^{1+\alpha_t} \ldots p_k^{\alpha_k} p^j & \text{if } p_+^j = p_t^{1+\alpha_t} \\[2mm] p_1^{\alpha_1} \ldots p_k^{\alpha_k} p^j q & \text{if } p_+^j = q. \end{cases}
$$

By using that $\Phi(p^{-j}) = \frac{p}{\Phi(p^j)}$ and $p_-^j = \left(p_+^j\right)^{-1}$ we have

$$
\Phi(p_-^j) = \Phi\left(\left(p_+^j\right)^{-1}\right) = \begin{cases} \dfrac{p_t}{\Phi\left(p_+^j\right)} & \text{if } p_+^j = p_t^{1+\alpha_t} \\[4mm] \dfrac{q}{\Phi\left(p_+^j\right)} & \text{if } p_+^j = q, \end{cases}
$$

which implies that $\Phi(p_-^j) = \frac{1}{p_1^{\alpha_1} \ldots p_k^{\alpha_k} p^j} = \frac{1}{\Phi(p^j)}$, and thus $p\Phi(p_-^j) = \frac{p}{\Phi(p^j)} = \Phi\left(p^{-j}\right)$, i.e. $\Phi(p_-^j) = \frac{\Phi(p^{-j})}{p}$. ∎

Lemma 76

(i) *The adelic ball $B_r := B_r(0)$ is a compact subset and its volume is given by*

$$
\mathrm{vol}\,(B_r) = \Phi(r).
$$

(ii) *The adelic sphere $S_r := S_r(0)$ is a compact subset and its volume is given by*

$$
\mathrm{vol}\,(S_r) = \Phi(r) - \Phi(r_-).
$$

Proof The compactness of $B_r(0)$ was established in Corollary 72. Since $S_r(0)$ is a closed subset of $\mathbb{A}_f$ and $S_r(0) \subset B_r(0)$ we conclude that $S_r(0)$ is compact. The formulas for volumes follows immediately from (4.9), (4.11) and (4.7). ∎

4.3.2   The Fourier Transform of Radial Functions

Definition 77 A function $f : \mathbb{A}_f \to \mathbb{C}$ is said to be radial if its restriction to any sphere S_r, $r > 0$, is a constant function, i.e. $f|_{S_r} = f_r \in \mathbb{C}$, $r > 0$.

If f is a radial function, then there exists a function $h : \mathbb{R} \to \mathbb{C}$ such that $f(\xi) = h(\|x\|)$. By abuse of notation, we will denote a radial function f in the form $f = f(\|x\|)$.

Lemma 78 *Let $f : \mathbb{A}_f \to \mathbb{C}$ be an integrable function. Then the following assertions hold:*

(i)

$$\int_{\mathbb{A}_f} f(x)\, dx_{\mathbb{A}_f} = \sum_{p^m, m \in \mathbb{Z} \setminus \{0\}} \int_{S_{p^m}} f(x)\, dx_{\mathbb{A}_f}.$$

In the particular case in which f is a radial function this formula takes the form

$$\int_{\mathbb{A}_f} f(x)\, dx_{\mathbb{A}_f} = \sum_{p^m, m \in \mathbb{Z} \setminus \{0\}} f(p^m)\, \mathrm{vol}\left(S_{p^m}\right).$$

(ii) Take $A^{(i)} = \bigsqcup_{m \in J} S_{p^m} \subset \mathbb{A}_f$, where J is a (countable) subset of $\mathbb{Z} \setminus \{0\}$, then

$$\int_{\mathbb{A}_f} f(x)\, 1_{A^{(i)}}(x)\, dx_{\mathbb{A}_f} = \sum_{p^m, m \in J} \int_{S_{p^m}} f(x)\, dx_{\mathbb{A}_f}.$$

In the particular case in which f is a radial function this formula takes the form

$$\int_{\mathbb{A}_f} f(x)\, 1_{A^{(i)}}(x)\, dx_{\mathbb{A}_f} = \sum_{p^m, m \in J} f(p^m)\, \mathrm{vol}\left(S_{p^m}\right).$$

(iii) Assume that $\mathbb{A}_f = \bigsqcup_{i \in \mathbb{N}} A^{(i)}$ with each $A^{(i)}$ is a disjoint union of spheres, then

$$\int_{\mathbb{A}_f} f(x)\, dx_{\mathbb{A}_f} = \sum_{i \in \mathbb{N}} \int_{A^{(i)}} f(x)\, dx_{\mathbb{A}_f}.$$

Proof The proof follows by general techniques in measure theory, the compactness of the adelic balls and spheres, see Lemma 76, and the characterization of the adelic integrals for positive functions given in [50, p. 21]. ∎

To simplify notations, throughout this subsection the expressions $\|0\|^{-1}$ and $|0|_p^{-1}$ in the inequalities mean ∞. The following theorem describes the Fourier transform of a radial function.

Theorem 79 *Let $f = f(\|x\|) : \mathbb{A}_f \to \mathbb{C}$ be a radial function in $L^1(\mathbb{A}_f)$. Then the following formula holds:*

$$\hat{f}(\xi) = \sum_{q^j < \|\xi\|^{-1}} \Phi\left(q^j\right) \left(f(q^j) - f(q^j_+)\right) \quad \text{for any } \xi \in \mathbb{A}_f, \tag{4.16}$$

where q^j runs through all non-zero powers of prime numbers; the functions $\|\xi\|$, $\Phi(\xi)$ and q^j_+ are defined by (4.5), (4.11) and (4.12).

Remark 80 It follows from (4.16) that the Fourier transform of a radial function is again a radial function.

Proof We represent the ring of finite adeles $\mathbb{A}_f$ as a disjoint union of the following sets

$$\mathbb{A}_f = \{0\} \sqcup \bigsqcup_q \mathbb{A}^{(0,q)} \sqcup \bigsqcup_q \mathbb{A}^{(1,q)},$$

where

$$\mathbb{A}^{(0,q)} := \bigsqcup_{j<0} S_{q^j} = \left\{x \in \mathbb{A}_f; 0 < \|x\| < 1, \ \|x\| = \frac{|x_q|_q}{q} \text{ and } \frac{|x_p|_p}{p} < \frac{|x_q|_q}{q} \text{ for } p \neq q\right\},$$

$$\mathbb{A}^{(1,q)} := \bigsqcup_{j>0} S_{q^j} = \left\{x \in \mathbb{A}_f; \|x\| > 1, \ \|x\| = |x_q|_q \text{ and } |x_p|_p < |x_q|_q \text{ for } p \neq q\right\}.$$

Note that on the sets $\mathbb{A}^{(0,q)}$ we have $\|x\| = \|x\|_0$ and on the sets $\mathbb{A}^{(1,q)}$ we have $\|x\| = \|x\|_1$. Then $\hat{f}(\xi) = \sum_q \hat{f}^{(0,q)}(\xi) + \sum_q \hat{f}^{(1,q)}(\xi)$, where

$$\hat{f}^{(k,q)}(\xi) := \int_{\mathbb{A}^{(k,q)}} \chi(\xi x) f(\|x\|) \, dx_{\mathbb{A}_f}, \qquad k = 0, 1, \ q \text{ is a prime.}$$

We set

$$\beta_q := \beta_q(\xi) = -[\log_q \|\xi\|] \tag{4.17}$$

with convention that $\beta_q(0) = +\infty$. We also set $\delta(t) = 1$ if $t = 0$ and $\delta(t) = 0$ otherwise.

To simplify the proof, we first present the final formulas for the functions $\hat{f}^{(k,q)}(\xi)$, the proofs are given later.

Claim 81

$$\sum_q \hat{f}^{(1,q)}(\xi) = 0 \qquad \text{if } \|\xi\| > 1, \tag{A}$$

$$\sum_q \hat{f}^{(1,q)}(\xi) = \sum_{q < \|\xi\|^{-1}} \left\{ \left(1 - \frac{1}{q}\right)^{\beta_q(x)-1} \sum_{j=1} f(q^j) \Phi(q^j) \right\}$$

(B)

$$- \sum_q \frac{1}{q} f(\|\xi\|^{-1}) \Phi(\|\xi\|^{-1}) \delta\left(\frac{|\xi_q|_q}{q} - \|\xi\|\right) \quad \text{if } \|\xi\| < 1.$$

Claim 82

$$\sum_q \hat{f}^{(0,q)}(\xi) = \sum_q \left\{ \left(1 - \frac{1}{q}\right)^{-1} \sum_{j=-\infty} f(q^j) \Phi(q^j) \right\} \quad \text{if } \|\xi\| < 1,$$

(C)

$$\sum_q \hat{f}^{(0,q)}(\xi) = \sum_q \left\{ \left(1 - \frac{1}{q}\right)^{\beta_q(\xi)-1} \sum_{j=-\infty} f(q^j) \Phi(q^j) \right\}$$

(D)

$$- \sum_q \frac{1}{q} f(\|\xi\|^{-1}) \Phi(\|\xi\|^{-1}) \delta(|\xi_q|_q - \|\xi\|) \quad \text{if } \|\xi\| > 1.$$

Combining (A), (B), (C), (D) we obtain

$$\hat{f}(\xi) = \sum_q \left\{ \left(1 - \frac{1}{q}\right) \sum_{\substack{j \leq \beta_q(\xi)-1. \\ j \neq 0}} f(q^j) \Phi(q^j) \right\}$$

(4.18)

$$- \sum_{q,j} \frac{1}{q} f(\|\xi\|^{-1}) \Phi(\|\xi\|^{-1}) \delta(q^j - \|\xi\|).$$

Note that the last sum over q and j involving the function δ means that we take the only term corresponding to the prime number q such that $\|\xi\| = q^j$ for some $j \in \mathbb{Z} \setminus \{0\}$.

Now the proof of the theorem may be finished as follows. Since $f \in L^1(\mathbb{A}_f)$ and $\mathrm{vol}(S_{p^j}) = \mathrm{vol}(\{x \in \mathbb{A}_f; \|x\| = p^j\}) = \Phi(p^j) - \Phi(p^j_-)$, see Lemma 76, the series $\sum_{q^j} (\Phi(q^j) - \Phi(q^j_-)) |f(q^j)|$ is convergent. Because of the inequality $\Phi(q^j) - \Phi(q^j_-) \geq \frac{1}{2}\Phi(q^j)$ the series $\sum_{q^j} \Phi(q^j) |f(q^j)|$ converges as well, hence we may arbitrary reorder the terms in (4.18).

By the properties of the entire part function, the following inequalities hold for the function β_q (see (4.17)):

$$q^j \leq q^{-\lceil \log_q \|\xi\|\rceil - 1} < q^{-\log_q \|\xi\|} = \|\xi\|^{-1}, \qquad j < \beta_q, \ j \in \mathbb{Z}, \qquad (4.19)$$

$$q^j \geq q^{-\lceil \log_q \|\xi\|\rceil} \geq q^{-\log_q \|\xi\|} = \|\xi\|^{-1}, \qquad j \geq \beta_q, \ j \in \mathbb{Z}, \qquad (4.20)$$

where the equality in the second inequality is possible only when $\|\xi\|$ is a power of q. Suppose in (4.18), that $\|\xi\| = p^k$ for some prime number p and integer $k \neq 0$. It follows from inequalities (4.19), (4.20) that the formula (4.18) may be written as

$$
\begin{aligned}
\hat{f}(\xi) &= \sideset{}{'}\sum_{q^j < \|\xi\|^{-1}} \left(1 - \frac{1}{q}\right) f(q^j)\Phi(q^j) - \frac{1}{p} f\left(\|\xi\|^{-1}\right)\Phi\left(\|\xi\|^{-1}\right) \\
&= \sideset{}{'}\sum_{q^j < \|\xi\|^{-1}} f(q^j)\Phi(q^j) - \sideset{}{'}\sum_{q^j < \|\xi\|^{-1}} f(q^j)\Phi(q_-^j) - f\left(\|\xi\|^{-1}\right)\Phi\left(\|\xi\|_-^{-1}\right) \\
&= \sideset{}{'}\sum_{q^j < \|\xi\|^{-1}} f(q^j)\Phi(q^j) - \sideset{}{'}\sum_{q^j \leq \|\xi\|^{-1}} f(q^j)\Phi(q_-^j) \\
&= \sideset{}{'}\sum_{q^j < \|\xi\|^{-1}} \Phi(q^j)\left(f(q^j) - f(q_+^j)\right),
\end{aligned}
$$

where we have used (4.12)–(4.15) and $\sum'$ means that the value $j = 0$ is omitted in the summation.

The checking of the formula (4.16) in the case $\|\xi\| = 0$ is left to the reader. ∎

Proof of Claim 81 We assume that $\xi \neq 0$. The case $\xi = 0$ may be checked directly. With the use of (4.2), Lemma 78 and the fact that $1_{S_r}(x)$ is a factorizable function, we may write $\hat{f}^{(1,q)}$ as

$$
\hat{f}^{(1,q)}(\xi) = \int_{|x_q|_q \geq q} \chi_q\left(x_q \xi_q\right) f(|x_q|_q) \left\{ \prod_{p \neq q} \int_{|x_p|_p < |x_q|_q} \chi_p\left(x_p \xi_p\right) dx_p \right\} dx_q.
$$

Denote by $\alpha_p(x_q)$ the largest integer satisfying $p^{\alpha_p(x_q)} \leq |x_q|_q$ (i.e. $\alpha_p(x_q) = [\log_p |x_q|_q] = [[\log_p |x_q|_q]]$). Note that the equality $p^{\alpha_p(x_q)} = |x_q|_q$ is impossible for $|x_q|_q > 1$ and $p \neq q$, hence $p^{\alpha_p(x_q)} < |x_q|_q$. Recall that

$$
\int_{|x_p|_p \leq p^{\alpha_p(x_q)}} \chi_p\left(x_p \xi_p\right) dx_p = \begin{cases} p^{\alpha_p(x_q)} & \text{if } |\xi_p|_p \leq p^{-\alpha_p(x_q)}, \\ 0 & \text{if } |\xi_p|_p \geq p^{-\alpha_p(x_q)+1}, \end{cases}
$$

and since $p^{\alpha_p(x_q)} < |x_q|_q < p^{\alpha_p(x_q)+1}$, we have

$$
\int_{|x_p|_p \leq p^{\alpha_p(x_q)}} \chi_p\left(x_p \xi_p\right) dx_p = \begin{cases} p^{\alpha_p(x_q)} & \text{if } |\xi_p|_p < p|x_q|_q^{-1}, \\ 0 & \text{if } |\xi_p|_p > p|x_q|_q^{-1}, \end{cases}
$$

which implies

$$\prod_{p\neq q}\int_{|x_p|_p<|x_q|_q}\chi_p\left(x_p\xi_p\right)dx_p=\left(\prod_{p\neq q}p^{\alpha_p(x_q)}\right)1_B\left(x_q\right)=\frac{\Phi(|x_q|_q)}{|x_q|_q}1_B\left(x_q\right),$$

(4.21)

where $1_B\left(x_q\right)$ is the characteristic function of the set

$$B:=\left\{x_q\in\mathbb{Q}_q;\max_{p\neq q}\frac{|\xi_p|_p}{p}<|x_q|_q^{-1}\right\}.$$

Therefore

$$\hat{f}^{(1,q)}\left(\xi\right)=\int_{q\leq|x_q|_q<\left(\max_{p\neq q}\frac{|\xi_p|_p}{p}\right)^{-1}}\chi_q\left(x_q\xi_q\right)f(|x_q|_q)\frac{\Phi(|x_q|_q)}{|x_q|_q}dx_q.$$

(4.22)

Note that it follows from (4.22) that $\hat{f}^{(1,q)}\left(\xi\right)=0$ if $\max_{p\neq q}\frac{|\xi_p|_p}{p}\geq\frac{1}{q}$.

Set γ_q to be the largest integer satisfying $q^{\gamma_q}<\left(\max_{p\neq q}\frac{|\xi_p|_p}{p}\right)^{-1}$, then

$$\hat{f}^{(1,q)}\left(\xi\right)=\sum_{j=1}^{\gamma_q}\frac{f(q^j)\Phi\left(q^j\right)}{q^j}\int_{|x_q|_q=q^j}\chi_q\left(x_q\xi_q\right)dx_q.$$

(4.23)

We recall that

$$\int_{|x_q|_q=q^j}\chi_q\left(x_q\xi_q\right)dx_q=\begin{cases}q^j\left(1-q^{-1}\right) & \text{if }|\xi_q|_q\leq q^{-j},\\-q^{j-1} & \text{if }|\xi_q|_q=q^{-j+1},\\0 & \text{if }|\xi_q|_q\geq q^{-j+2}.\end{cases}$$

(4.24)

Note that the integral (4.24) is non-zero when $|x_q|_q\leq\frac{q}{|\xi_q|_q}=\left(\frac{|\xi_q|_q}{q}\right)^{-1}$. Since for $|x_q|_q>1$ the equality $|x_q|_q=\left(\max_{p\neq q}\frac{|\xi_p|_p}{p}\right)^{-1}$ is impossible, the last inequality may be combined with $|x_q|_q<\left(\max_{p\neq q}\frac{|\xi_p|_p}{p}\right)^{-1}$ into the inequality $|x_q|_q\leq\left(\max_p\frac{|\xi_p|_p}{p}\right)^{-1}=\|\xi\|_0^{-1}$, cf. (4.3). Then it follows from (4.22)–(4.24) that

$$\hat{f}^{(1,q)}\left(\xi\right)=\int_{q\leq|x_q|_q\leq\|\xi\|_0^{-1}}\frac{\chi_q\left(x_q\xi_q\right)f(|x_q|_q)\Phi(|x_q|_q)}{|x_q|_q}dx_q.$$

(4.25)

Note that $\hat{f}^{(1,q)}(\xi) = 0$ if $\|\xi\|_0^{-1} < q$ and that

$$\left\{\xi \in \mathbb{A}_f; \|\xi\|_0^{-1} \geq 2\right\} = \left\{\xi \in \mathbb{A}_f; \max_p \frac{|\xi_p|_p}{p} \leq \frac{1}{2}\right\} = \prod_p \mathbb{Z}_p,$$

hence for any ξ outside of $\prod_p \mathbb{Z}_p$ the sum $\sum_q \hat{f}^{(1,q)}(\xi)$ vanishes. Therefore we have non-zero terms in $\sum_q \hat{f}^{(1,q)}(\xi)$ only if $\|\xi\|_0 < 1$. In such case $\|\xi\|_0 = \|\xi\|$. We recall definition (4.17) of β_q and inequalities (4.19), (4.20). Then it follows from (4.23)–(4.25) that

$$\hat{f}^{(1,q)}(\xi) = \left(1 - \frac{1}{q}\right) \sum_{j=1}^{\beta_q - 1} f(q^j)\Phi(q^j) \qquad \text{if } \frac{|\xi_q|_q}{q} < \|\xi\| \tag{4.26}$$

and

$$\hat{f}^{(1,q)}(\xi) = \left(1 - \frac{1}{q}\right) \sum_{j=1}^{\beta_q - 1} f(q^j)\Phi(q^j) - \frac{1}{q}f\left(\|\xi\|^{-1}\right)\Phi\left(\|\xi\|^{-1}\right) \qquad \text{if } \frac{|\xi_q|_q}{q} = \|\xi\|. \tag{4.27}$$

By combining the formulas (4.26)–(4.27) we obtain formula (B).   ∎

Proof of Claim 82 We assume that $\xi \neq 0$. The case $\xi = 0$ may be checked directly. The required calculations are mostly similar to the previous ones, however there are some subtle variations. We have $\|x\| = q^{-1}|x_q|_q$ and

$$\hat{f}^{(0,q)}(\xi) = \int_{q^{-1}|x_q|_q < 1} \chi_q\left(x_q\xi_q\right)f(q^{-1}|x_q|_q)\left\{\prod_{p \neq q} \int_{p^{-1}|x_p|_p < q^{-1}|x_q|_q} \chi_p\left(x_p\xi_p\right)dx_p\right\}dx_q.$$

Let $\alpha_p(x_q)$ denote the largest power $p^{\alpha_p(x_q)}$ satisfying $p^{-1}p^{\alpha_p(x_q)} < q^{-1}|x_q|_q$ which is equal to $1 + \left[\log_p q^{-1}|x_q|_q\right]$, and since $q^{-1}|x_q|_q < 1$, the last quantity is equal to $\left[\left[\log_p q^{-1}|x_q|_q\right]\right]$, cf. (4.7). Hence similarly to (4.21) with the use of (4.11) we obtain

$$\prod_{p \neq q} \int_{p^{-1}|x_p|_p < q^{-1}|x_q|_q} \chi_p\left(x_p\xi_p\right)dx_p = \left(\prod_{p \neq q} p^{\alpha_p(x_q)}\right)1_B\left(x_q\right)$$

$$= \frac{\Phi(q^{-1}|x_q|_q)}{q^{\left[\left[\log_q q^{-1}|x_q|_q\right]\right]}}1_B\left(x_q\right),$$

where $1_B\left(x_q\right)$ is the characteristic function of the set

$$B := \left\{x_q \in \mathbb{Q}_q; |x_q|_q < q\left(\max_{p \neq q}|\xi_p|_p\right)^{-1}\right\}. \tag{4.28}$$

Since $|x_q|_q$ is a power of q and $q^{-1}|x_q|_q < 1$, we have $[[\log_q q^{-1}|x_q|_q]] = \log_q |x_q|_q$ and $q^{[[\log_q q^{-1}|x_q|_q]]} = q^{\log_q |x_q|_q} = |x_q|_q$. Since $q^{-1}|x_q|_q < 1$ is equivalent to $|x_q|_q \leq 1$, we obtain

$$\hat{f}^{(0,q)}\left(\xi\right) = \int\limits_{\substack{|x_q|_q \leq 1. \\ |x_q|_q < q(\max_{p \neq q}|\xi_p|_p)^{-1}}} \chi_q\left(x_q\xi_q\right) f(q^{-1}|x_q|_q) \frac{\Phi(q^{-1}|x_q|_q)}{|x_q|_q}dx_q. \tag{4.29}$$

It follows from (4.24) that the last integral is non-zero when $|\xi_q|_q \leq q|x_q|_q^{-1}$, which may be combined with (4.28) into $\max_p |\xi_p|_p \leq q|x_q|_q^{-1}$ or, equivalently, into $|x_q|_q \leq q(\max_p |\xi_p|_p)^{-1}$. Hence the domain of integration in the last integral is

$$|x_q|_q \leq \min\left\{1, \frac{q}{\max_p |\xi_p|_p}\right\}, \tag{4.30}$$

and we have to consider two cases.

Case 1 $\|\xi\| < 1$, i.e. $\max_p |\xi_p|_p \leq 1$. In this case the minimum in (4.30) is equal to 1 and the inequality $|\xi_q|_q < q|x_q|_q^{-1}$ always holds, hence from (4.29), (4.30) and (4.24) similarly to (4.26) we obtain

$$\hat{f}^{(0,q)}\left(\xi\right) = \left(1 - \frac{1}{q}\right)\sum_{j=-\infty}^{0} f(q^{j-1})\Phi(q^{j-1}) = \left(1 - \frac{1}{q}\right)\sum_{j=-\infty}^{-1} f(q^{j})\Phi(q^{j}).$$

Case 2 $\|\xi\| > 1$, i.e. $\max_p |\xi_p|_p = \|\xi\| > 1$. In this case it is sufficient to determine possible values of $|x_q|_q$ from the inequality $|x_q|_q \leq \frac{q}{\|\xi\|}$, because they satisfy the inequality $|x_q|_q \leq 1$ even in the case $\|\xi\| < q$. Recall definition (4.17) of β_q and inequalities (4.19), (4.20). As a result, from (4.29), (4.30) and (4.24) similarly to (4.26) and (4.27) we have

$$\hat{f}^{(0,q)}\left(\xi\right) = \left(1 - \frac{1}{q}\right)\sum_{j=-\infty}^{\beta_q} f(q^{j-1})\Phi(q^{j-1})$$

$$= \left(1 - \frac{1}{q}\right)\sum_{j=-\infty}^{\beta_q-1} f(q^{j})\Phi(q^{j}) \qquad \text{if } |\xi_q|_q < \|\xi\|$$

and

$$\hat{f}^{(0,q)}(\xi) = \left(1 - \frac{1}{q}\right) \sum_{j=-\infty}^{\beta_q-1} f(q^j)\Phi(q^j) - \frac{1}{q}f\left(\|\xi\|^{-1}\right)\Phi\left(\|\xi\|^{-1}\right) \quad \text{if } |\xi_q|_q = \|\xi\|.$$

The last two equalities give us formulas (C)–(D).   ∎

As a corollary from Theorem 79 we derive a sufficient condition for the Fourier transform of a radial function to be non-negative.

Corollary 83 *Let f be a real-valued non-increasing radial function, i.e. $f = f(\|x\|)$ and $f(x) \geq f(y)$ for any $x, y \in \mathbb{A}_f$ satisfying $\|x\| \leq \|y\|$. Then*

$$\hat{f}(\xi) \geq 0 \qquad \text{for any } \xi \in \mathbb{A}_f.$$

On the base of Theorem 79 we may compute the Fourier transforms of characteristic functions of balls and spheres.

Corollary 84 *Let f be a characteristic function of a ball, i.e. $f = 1_{B_r}(x)$, with r in $\{p^j; p \text{ is prime}, j \in \mathbb{Z} \setminus \{0\}\}$. Then*

$$\hat{f}(\xi) = \Phi(r)1_{B_R}(\xi), \quad R = (r^{-1})_-.$$

Let g be a characteristic function of a sphere, i.e. $g = 1_{S_r}(x)$, with r in $\{p^j; p \text{ is prime}, j \in \mathbb{Z} \setminus \{0\}\}$. Then

$$\hat{g}(\xi) = \Phi(r)1_{B_R}(\xi) - \Phi(r_-)1_{B_{R_+}}(\xi), \quad R = (r^{-1})_-.$$

Proof As it follows from (4.16) we have at most one non-zero term equal to $\Phi(r)$ in the Fourier transform of the ball B_r, and this term is present if and only if $r < \|x\|^{-1}$ which is equivalent to $\|x\| < r^{-1}$ or $\|x\| \leq (r^{-1})_-$.

The second statement follows from the presentation $S_r = B_r \setminus B_{r_-}$ and the properties of the operators '−' and '+': $((r_-)^{-1})_- = ((r^{-1})_+)_- = r^{-1} = R_+$.   ∎

4.3.3 Distributions on $\mathbb{A}_f$

In this subsection we consider $\mathbb{A}_f$ as the complete metric space $(\mathbb{A}_f, \rho)$. As it was previously mentioned, the space $\mathcal{D}(\mathbb{A}_f)$ of Bruhat-Schwartz functions consists of finite linear combinations of factorizable functions $f = \prod_p f_p$, where a finite number of the functions f_p are in $\mathcal{D}(\mathbb{Q}_p)$ and the rest of the functions are the characteristic functions of the sets $\mathbb{Z}_p$, i.e. $f = \prod_{p \leq N} f_p \times \prod_{p > N} \Omega_p(|x_p|_p)$. For the sake of simplicity, from now on we will use *test function* to mean *Bruhat-Schwartz function*. The spaces $\mathcal{D}(\mathbb{Q}_p)$ consist of compactly supported locally constant functions. We show that the same property characterizes the space $\mathcal{D}(\mathbb{A}_f)$. Despite a similar result was already proved in [72], the adelic metric ρ allows us to introduce the notion of 'parameter of constancy' for functions in $\mathcal{D}(\mathbb{A}_f)$ and to give a construction of a topology for $\mathcal{D}(\mathbb{A}_f)$ in a similar way as for the spaces $\mathcal{D}(\mathbb{Q}_p)$.

Definition 85 We say that a function f is locally constant if for any $x \in \mathbb{A}_f$ there exists a constant $\ell(x) > 0$ such that $f(x + y) = f(x)$ for any $y \in B_{\ell(x)}(0)$.

The same reasoning as in the p-adic case, see e.g. [111] or [5], shows that for a compactly supported function f the same constant ℓ may be chosen for all points $x \in \mathbb{A}_f$.

Definition 86 Let f be a non-zero compactly supported function. We define the parameter of constancy ℓ of f as the largest non-zero integer power of a prime number such that

$$f(x + y) = f(x) \qquad \text{for any } x \in \mathbb{A}_f, \ y \in B_\ell(0). \tag{4.31}$$

By definition we set the parameter of constancy of function 0 to be equal $+\infty$.

Note that the above definition induces naturally a notion of parameter of constancy for functions on $\mathbb{Q}_p$ which is different from the one presented in Chap. 1 and [111]. Indeed, the parameter of local constancy belongs to $\mathbb{N} \cup \{+\infty\}$, while in Chap. 1, the parameter of parameter of local constancy belongs to $\mathbb{Z} \cup \{+\infty\}$.

Lemma 87 *The function $f \in \mathcal{D}(\mathbb{A}_f)$ if and only if it is locally constant with compact support.*

Proof The statement is trivial for $f \equiv 0$. Suppose $f \in \mathcal{D}(\mathbb{A}_f) \setminus \{0\}$, and $f = \sum_{m=1}^{M} f^{(m)}(x)$, where each function $f^{(m)}$ is factorizable, $f^{(m)} = \prod_{p \leq N_m} f_p^{(m)} \times \prod_{p > N_m} \Omega_p(|x_p|_p)$. Since each $f^{(m)}$ is compactly supported, so is f.

Let $l_p^{(m)}$ denote the parameter of constancy of the function $f_p^{(m)}$, i.e. $f_p^{(m)}(x_p + y_p) = f_p^{(m)}(x_p)$ for all y_p such that $|y_p|_p \leq \ell_p^{(m)}$. Consider

$$\ell = \min \left\{ \frac{1}{2}, \min_{p,m} \frac{\ell_p^{(m)}}{p} \right\}.$$

Since we have only finite number of parameters $\ell_p^{(m)}$, $\ell > 0$. It is easy to check that (4.31) holds with this parameter ℓ, i.e. the function f is locally constant.

Suppose now that f is a locally constant function with compact support. Let $K = \operatorname{supp} f$. Since f is locally constant, for each $x \in K$ there exists a ball $B_{r(x)}(x)$, which is an open set, such that f is constant on $B_{r(x)}(x)$. Then there exists a finite number of these balls, say $B_{r_1}(x_1), \ldots, B_{r_n}(x_n)$, covering K. Since the metric is non-Archimedean we may assume that these balls are disjoint. Therefore

$$f(x) = f(x_1) \cdot 1_{B_{r_1}(x_1)}(x) + \ldots + f(x_n) \cdot 1_{B_{r_n}(x_n)}(x), \tag{4.32}$$

where each characteristic function $1_{B_r}(x)$ is factorizable, cf. (4.9).   ∎

Remark 88 Let $\mathcal{P}(\mathbb{A}_f)$ denote the set of parameters of constancy of functions from $\mathcal{D}(\mathbb{A}_f)$. Then

$$\mathcal{P}(\mathbb{A}_f) = \{\ell \in \mathbb{Q}; \ell = p^m, \ p \text{ is a prime}, \ m \in \mathbb{Z} \setminus \{0\}\} \cup \{+\infty\}.$$

By considering the characteristic functions of the adelic balls B_r we verify that every number in $\mathcal{P}(\mathbb{A}_f)$ is an admissible parameter of constancy. $\mathcal{P}(\mathbb{A}_f)$ is a countable and totally ordered set.

We define by $\mathcal{D}_R^\ell(\mathbb{A}_f)$ the subspace of test functions with supports contained in the adelic ball B_R and parameters of constancy $\geq \ell$. Then the following embedding holds: $\mathcal{D}_R^\ell(\mathbb{A}_f) \subset \mathcal{D}_{R'}^{\ell'}(\mathbb{A}_f)$ whenever $R \leq R', \ell \geq \ell'$. As in the p-adic setting, see e.g. [5, 105, 111], we define the convergence in $\mathcal{D}(\mathbb{A}_f)$ in the following way: $f_k \to 0$, $k \to \infty$ in $\mathcal{D}(\mathbb{A}_f)$ if and only if

(i) $f_k \in \mathcal{D}_R^\ell(\mathbb{A}_f)$ where R and ℓ do not depend on k;
(ii) $f_k \to 0$ uniformly as $k \to \infty$.

With this notion of convergence $\mathcal{D}(\mathbb{A}_f)$ becomes a complete topological vector space. In addition,

$$\mathcal{D}_R(\mathbb{A}_f) = \lim_{\ell \to 0} \text{ind } \mathcal{D}_R^\ell(\mathbb{A}_f), \qquad \mathcal{D}(\mathbb{A}_f) = \lim_{R \to \infty} \text{ind } \mathcal{D}_R(\mathbb{A}_f).$$

Note that the second inductive limit makes sense because $\mathcal{P}(\mathbb{A}_f)$ is totally ordered.

The following proposition shows that the spaces $\mathcal{D}_R^\ell(\mathbb{A}_f)$ possess similar properties to their p-adic analogues.

Proposition 89 *For arbitrary $\ell \leq R$ the space $\mathcal{D}_R^\ell(\mathbb{A}_f)$ is non-trivial and finite dimensional, its dimension is equal to $\Phi(R)/\Phi(\ell)$, with a basis given by the characteristic functions of disjoint balls $B_\ell(x^{(n)}) \subset B_R$. If $f \in \mathcal{D}_R^\ell(\mathbb{A}_f)$ then $\hat{f}(\xi) = \mathcal{F}_{x \to \xi} f \in \mathcal{D}_{(1/\ell)_-}^{(1/R)_-}(\mathbb{A}_f)$. Moreover, $\mathcal{F}\mathcal{D}_R^\ell(\mathbb{A}_f) = \mathcal{D}_{(1/\ell)_-}^{(1/R)_-}(\mathbb{A}_f)$.*

Proof Note that B_R is a finite disjoint union of balls of type $B_\ell(x_i)$ and the number of such balls is $\text{vol}(B_R)/\text{vol}(B_\ell)$. The first statement follows from this observation by (4.32).

For the second part it is enough to consider the Fourier transform of the characteristic function of a ball $B_\ell(x^{(n)}) \subset B_R$. We obtain from Corollary 84

$$\hat{1}_{B_\ell(x^{(n)})}(\xi) = \int_{\mathbb{A}_f} \chi(x\xi) 1_{B_\ell}(x - x^{(n)}) \, dx_{\mathbb{A}_f}$$

$$= \chi(x^{(n)}\xi) \hat{1}_{B_\ell}(\xi) = \chi(x^{(n)}\xi) \Phi(\ell) 1_{B_{(1/\ell)_-}}(\xi),$$

hence the Fourier transform is supported in the ball $B_{(1/\ell)_-}$. Since $x^{(n)} \in B_R$, for any $y \in B_{(1/R)_-}$ we have $\|x^{(n)} \cdot y\| < 1$ hence $\chi(x^{(n)}y) = 1$ and the Fourier transform of a ball $B_\ell(x^{(n)})$ is locally constant with the parameter of constancy $\geq (1/R)_-$. The last part follows from the observation that $\left(\left((n^{-1})_-\right)^{-1}\right)_- = \left(\left((n^{-1})^{-1}\right)_+\right)_- = (n_+)_- = n$ for any non-zero power of a prime.   ∎

Proposition 90 *(i) Let K be a compact subset of $\mathcal{D}(\mathbb{A}_f)$. The space of test functions $\mathcal{D}(\mathbb{A}_f)$ is dense in the space $C(K)$ of continuous functions on K. (ii) The space of test functions $\mathcal{D}(\mathbb{A}_f)$ is dense in $L^\varrho\left(\mathbb{A}_f\right)$ for $1 \leq \varrho < \infty$.*

Proof The proof follows the classical pattern, see e.g. [5, 105, 111].   ∎

Denote by $\mathcal{D}'(\mathbb{A}_f)$ the $\mathbb{C}$-vector space of all (complex-valued) linear continuous functionals on $\mathcal{D}(\mathbb{A}_f)$. This space is *the space of Bruhat-Schwartz distributions* on $\mathbb{A}_f$. For the sake of simplicity we will use *distribution* instead of *Bruhat-Schwartz distribution*. We equip $\mathcal{D}'(\mathbb{A}_f)$ with the weak topology. The following proposition allows to simplify checking that a functional belongs to the space $\mathcal{D}'(\mathbb{A}_f)$ stating that every linear functional on $\mathcal{D}(\mathbb{A}_f)$ is continuous.

Proposition 91 *(i) $\mathcal{D}'(\mathbb{A}_f)$ is the $\mathbb{C}$-vector space of all (complex-valued) linear functionals on $\mathcal{D}(\mathbb{A}_f)$. (ii) $\mathcal{D}'(\mathbb{A}_f)$ is complete.*

Proof Due to Proposition 89 the proof of this proposition is completely similar to the proof given for the analogous statement in the p -adic case, see e.g. [5, 111].   ∎

4.3.4 *Pseudodifferential Operators and the Lizorkin Space on* $\mathbb{A}_f$

As it was mentioned in the introduction, the classical derivative cannot be defined for complex-valued functions on adeles. Instead we consider pseudodifferential operators. The function $\| \cdot \|$ which generates the metric ρ allows us to introduce a natural generalization of the Taibleson operator $D_{\mathbb{A}_f}^\gamma =: D^\gamma$, $\gamma > 0$, defined on $\mathcal{D}(\mathbb{A}_f)$ by

$$(D^\gamma f)(x) = \mathcal{F}_{\xi \to x}^{-1}\left(\|\xi\|^\gamma \,\mathcal{F}_{x \to \xi} f\right), \qquad f \in \mathcal{D}(\mathbb{A}_f). \qquad (4.33)$$

Lemma 92 *With the above notation,*

$$D^\gamma : \mathcal{D}(\mathbb{A}_f) \to C\left(\mathbb{A}_f\right) \cap L^2\left(\mathbb{A}_f\right).$$

Proof Since $\mathcal{F}_{x \to \xi} f$ may be represented as a linear combination of functions of type $1_{B_r(\xi_0)}(\xi)$, it is sufficient to consider the case $\mathcal{F}_{x \to \xi} f = 1_{B_r(\xi_0)}(\xi)$. If $0 \notin B_r(\xi_0)$ then $\|\xi\|^\gamma \, 1_{B_r(\xi_0)}(\xi) \in \mathcal{D}(\mathbb{A}_f)$ because $\|\xi\|^\gamma$ is locally constant outside of the origin, and hence

$$\mathcal{F}_{\xi \to x}^{-1}\left(\|\xi\|^\gamma \, 1_{B_r(\xi_0)}(\xi)\right) \in \mathcal{D}(\mathbb{A}_f) \subset L^2\left(\mathbb{A}_f\right).$$

If $0 \in B_r(\xi_0)$ then $B_r(\xi_0) = B_r(0)$ and $\|\xi\|^\gamma \leq r^\gamma$ on the $B_r(0)$, hence $\|\xi\|^\gamma 1_{B_r(0)}(\xi) \in L^1(\mathbb{A}_f) \cap L^2(\mathbb{A}_f)$. Thus $\mathcal{F}^{-1}_{\xi \to x}(\|\xi\|^\gamma 1_{B_r(\xi_0)}(\xi)) \in C(\mathbb{A}_f) \cap L^2(\mathbb{A}_f)$. ∎

The space $\mathcal{D}(\mathbb{A}_f)$ is not invariant under the action of the operator D^γ. To overcome such an inconvenience, we introduce the following space

$$\mathcal{L}_0(\mathbb{A}_f) := \mathcal{L}_0 = \{f \in \mathcal{D}(\mathbb{A}_f); \hat{f}(0) = 0\}.$$

The space $\mathcal{L}_0$ can be equipped with the topology of the space $\mathcal{D}(\mathbb{A}_f)$, which makes $\mathcal{L}_0$ a complete space. Note that

$$\mathcal{L}_0 = \mathcal{F}\{h \in \mathcal{D}(\mathbb{A}_f); h(0) = 0\}. \tag{4.34}$$

This space is an adelic analogue of the Lizorkin space of the second kind. We refer the reader to [5] for the theory of the p-adic Lizorkin spaces. Recently in [74] an adelic version of the Lizorkin space of the first kind was introduced.

Lemma 93 *With the above notation the following assertions hold:*

(i) $D^\gamma \mathcal{L}_0 = \mathcal{L}_0$ *for* $\gamma > 0$.
(ii) $f \in \mathcal{L}_0$ *if and only if* $f \in \mathcal{D}$ *and* $\int_{\mathbb{A}_f} f(x)\, dx_{\mathbb{A}_f} = 0$.
(iii) $\mathcal{L}_0$ *is dense in* $\mathcal{D}$ *with respect to the* L^2-*norm.*
(iv) $\mathcal{L}_0$ *is dense in* $L^2(\mathbb{A}_f)$.

Proof

(i) Take $f \in \mathcal{L}_0$, then $\|\xi\|^\gamma \hat{f}(\xi) \in \mathcal{D}(\mathbb{A}_f)$ because $\hat{f}$ is equal to 0 in some neighborhood of 0 and $\|\xi\|^\gamma$ is locally constant outside of the origin. Therefore $D^\gamma f \in \mathcal{L}_0$, i.e. $D^\gamma \mathcal{L}_0 \subset \mathcal{L}_0$. The converse inclusion follows from the fact that $\frac{\hat{h}}{\|\xi\|^\gamma} \in \mathcal{D}(\mathbb{A}_f)$ for any $h \in \mathcal{L}_0$.
(ii) The statement follows from $\hat{f}(0) = \int_{\mathbb{A}_f} f(x)\, dx_{\mathbb{A}_f}$ which is the consequence of $\mathcal{F}\mathcal{D}(\mathbb{A}_f) = \mathcal{D}(\mathbb{A}_f)$.
(iii) By (4.34) and the fact that the Fourier transform preserves the L^2 norm, it is sufficient to show that $1_{B_r}(x)$ can be arbitrarily closely approximated by functions from $\mathcal{L}_0$ in the L^2 norm. As such approximating functions we may use $g_n(x) = 1_{B_r \setminus B_{p^{-n}}}(x)$.
(iv) The statement follows from (iii) since $\mathcal{D}(\mathbb{A}_f)$ is dense in $L^2(\mathbb{A}_f)$, see Proposition 90. ∎

We will use the notation $\mathcal{D}(D^\gamma)$ to denote the domain of the operator D^γ. We refer reader to [95] for notions of essentially self-adjoint operators and to [7] and [41] for the definition of strongly continuous (C_0) semigroups and related notions.

Since $\mathcal{L}_0(\mathbb{A}_f) \subset L^2(\mathbb{A}_f)$, we may consider the operator D^γ as an operator acting on $L^2(\mathbb{A}_f)$. It is easy to see that the operator D^γ with the domain $Dom(D^\gamma) = \mathcal{L}_0(\mathbb{A}_f)$ is symmetric. Moreover, similarly to the proof of Lemma 93 (i) we may check that $(\boldsymbol{D}^\gamma \pm i)\mathcal{L}_0(\mathbb{A}_f) = \mathcal{L}_0(\mathbb{A}_f)$, i.e. the ranges of the operators $\boldsymbol{D}^\gamma \pm i$ are dense in $L^2(\mathbb{A}_f)$, hence the operator $\boldsymbol{D}^\gamma$ is essentially self-adjoint, see [95, Corollary to Theorem VIII.3] for details. The following description of the self-adjoint closure holds.

Lemma 94 *The closure of the operator $\boldsymbol{D}^\gamma$, $\gamma > 0$ (let us denote it by $\boldsymbol{D}^\gamma$ again) with domain*

$$\mathrm{Dom}\,(\boldsymbol{D}^\gamma) := \left\{ f \in L^2\left(\mathbb{A}_f\right) ; \|\xi\|^\gamma \hat{f} \in L^2\left(\mathbb{A}_f\right) \right\} \tag{4.35}$$

is a self-adjoint operator. Moreover, the following assertions hold:

(i) $\boldsymbol{D}^\gamma$ is a positive operator;
(ii) $\boldsymbol{D}^\gamma$ is m-accretive, i.e. $-\boldsymbol{D}^\gamma$ is an m-dissipative operator;
(iii) the spectrum $\sigma(\boldsymbol{D}^\gamma) = \left\{ p^{\gamma j}; p \text{ is a prime}, j \in \mathbb{Z} \setminus \{0\} \right\} \cup \{0\};$
(iv) $-\boldsymbol{D}^\gamma$ is the infinitesimal generator of a contraction C_0 semigroup $\left(\mathcal{T}(t)\right)_{t \geq 0}$. Moreover, the semigroup $\left(\mathcal{T}(t)\right)_{t \geq 0}$ is bounded holomorphic (or analytic) with angle $\pi/2$.

Proof

(i) It follows from the Steklov–Parseval equality that for any $f \in L^2(\mathbb{A}_f)$

$$(\boldsymbol{D}^\gamma f, f) = (\|\xi\|^\gamma \mathcal{F}f, \mathcal{F}f) = \int_{\mathbb{A}_f} \|\xi\|^\gamma |\mathcal{F}f|^2 d\xi_{\mathbb{A}_f} \geq 0.$$

(ii), (iv) The result follows from the well-known corollary from the Lumer-Phillips Theorem, see e.g. [41, Chapter 2, Section 3] or [24]. For the property of the semigroup of being holomorphic, see e.g. [7, 3.7] or [41, Chapter 2, Section 4.7].

(iii) Since $\boldsymbol{D}^\gamma$ is self-adjoint and positive, $\sigma(\boldsymbol{D}^\gamma) \subset [0, \infty)$. Consider the eigenvalue problem $\boldsymbol{D}^\gamma f = \lambda f, f \in \mathcal{D}(\boldsymbol{D}^\gamma), \lambda > 0$. By applying the Fourier transform we obtain the equivalent equation

$$(\|\xi\|^\gamma - \lambda)\hat{f} = 0. \tag{4.36}$$

If $\lambda = p^{\gamma j}$ for some prime p and $j \in \mathbb{Z} \setminus \{0\}$ then the inverse Fourier transform of the characteristic function of $S_{p^j} = \{\xi \in \mathbb{A}_f; \|\xi\| = p^j\}$ is a solution of (4.36). If $\lambda \notin \{p^{\gamma j}; p \text{ is a prime}, j \in \mathbb{Z} \setminus \{0\}\}$, then the functions $\left|\frac{1}{\|\xi\|^\gamma - \lambda}\right|$ and $\left|\frac{\|\xi\|^\gamma}{\|\xi\|^\gamma - \lambda}\right|$ are bounded, hence the equation $\boldsymbol{D}^\gamma f - \lambda f = h$ is uniquely solvable for any $h \in L^2(\mathbb{A}_f)$ and $\lambda \in \rho(\boldsymbol{D}^\gamma)$. The point 0 belongs to $\sigma(\boldsymbol{D}^\gamma)$ as a limit point.   ∎

The representation of the generated semigroup $\left(\mathcal{T}(t)\right)_{t \geq 0}$ in the case $\gamma > 1$ is presented in detail in Theorem 114.

4.4 Metric Structures, Distributions and Pseudodifferential Operators on $\mathbb{A}$

4.4.1 A Structure of Complete Metric Space for the Adeles

We recall that $\mathbb{A} = \mathbb{R} \times \mathbb{A}_f$. Then any $x \in \mathbb{A}$ can be written uniquely as $x = (x_\infty, x_f) \in \mathbb{R} \times \mathbb{A}_f = \mathbb{A}$. Set for $x, y \in \mathbb{A}$

$$\rho_{\mathbb{A}}(x, y) := |x_\infty - y_\infty|_\infty + \rho(x_f, y_f),$$

where $\rho(x, y)$ was defined in (4.6). Then $(\mathbb{A}, \rho_{\mathbb{A}})$ is a complete metric space, see Proposition 71. Note that $\rho_{\mathbb{A}}(x, y)$ is topologically equivalent to

$$\tilde{\rho}(x, y) := \max\left\{|x_\infty - y_\infty|_\infty, \rho(x, y)\right\}, \qquad x, y \in \mathbb{A},$$

which induces on $\mathbb{A}$ the product topology. The topology of the restricted product on $\mathbb{A}$ is equal to the product topology on $\mathbb{R} \times \mathbb{A}_f$, where $\mathbb{R}$ is equipped with the usual topology and $\mathbb{A}_f$ with the restricted product topology. Hence the following result holds.

Proposition 95 *The restricted product topology on $\mathbb{A}$ is metrizable, a metric is given by $\rho_{\mathbb{A}}$. Furthermore, $(\mathbb{A}, \rho_{\mathbb{A}})$ is a complete metric space and $(\mathbb{A}, \rho_{\mathbb{A}})$ as a topological space is homeomorphic to $(\mathbb{R}, |\cdot|_\infty) \times (\mathbb{A}_f, \rho)$.*

Remark 96 $(\mathbb{A}, \rho_{\mathbb{A}})$ is a second-countable topological space. More precisely, $\left(B_\infty^{(i)} \times B_f^{(j)}\right)_{i,j \in \mathbb{N}}$ is a countable base, where $\left(B_\infty^{(i)}\right)_{i \in \mathbb{N}}$ is a countable base of $(\mathbb{R}, |\cdot|_\infty)$ and $\left(B_f^{(j)}\right)_{j \in \mathbb{N}}$ is a countable base of $(\mathbb{A}_f, \rho)$, see Remark 73 (ii). Therefore $(\mathbb{A}, \rho_{\mathbb{A}})$ is a semi-compact space.

4.4.2 Distributions on $\mathbb{A}$

The *space of Bruhat-Schwartz functions*, denoted $\mathcal{D}(\mathbb{A})$, consists of finite linear combinations of functions of type $h(x) = h_\infty(x_\infty) h_f(x_f)$ with $h_\infty \in \mathcal{D}(\mathbb{R})$, Schwartz space on $\mathbb{R}$, and $h_f \in \mathcal{D}(\mathbb{A}_f)$. The space $\mathcal{D}(\mathbb{A})$ is dense in $L^\varrho(\mathbb{A}, dx_{\mathbb{A}})$ for $1 \leq \varrho < +\infty$, see e.g. [37, Theorem 2.9]. The *space of distributions on $\mathcal{D}(\mathbb{A})$* is the strong dual space of $\mathcal{D}(\mathbb{A})$.

4.4.3   Pseudodifferential Operators and the Lizorkin Space on $\mathbb{A}$

We consider the pseudodifferential operator $D_{\mathbb{R}}^{\beta} =: D^{\beta}$, $\beta > 0$ on $\mathcal{D}(\mathbb{R})$ defined by

$$\left(D^{\beta}h\right)(x_{\infty}) = \mathcal{F}_{\xi_{\infty}\to x_{\infty}}^{-1}\left(|\xi_{\infty}|_{\infty}^{\beta}\mathcal{F}_{x_{\infty}\to\xi_{\infty}}h\right), \qquad h \in \mathcal{D}(\mathbb{R}). \tag{4.37}$$

Recall that the operator D^{β} is the real Riesz fractional operator and represents a fractional power of the Laplacian, see e.g. [101, §8], [102, §25].

We introduce the pseudodifferential operator $D_{\mathbb{A}}^{\alpha,\beta} =: D^{\alpha,\beta}$, $\alpha, \beta > 0$ on $\mathcal{D}(\mathbb{A})$ defined by

$$\left(D^{\alpha,\beta}h\right)(x) = \mathcal{F}_{\xi\to x}^{-1}\left((|\xi_{\infty}|_{\infty}^{\beta} + \|\xi_f\|^{\alpha})\mathcal{F}_{x\to\xi}h\right), \qquad h \in \mathcal{D}(\mathbb{A}). \tag{4.38}$$

Lemma 97 *With the above notation*

$$D^{\alpha,\beta} : \mathcal{D}(\mathbb{A}) \to C(\mathbb{A},\mathbb{C}) \cap L^{2}(\mathbb{A})$$

Proof It is sufficient to show the result for a factorizable function $h = h_{\infty}h_f$, $h_{\infty} \in \mathcal{D}(\mathbb{R})$, $h_f \in \mathcal{D}(\mathbb{A}_f)$. Since $\hat{h}(\xi) = \widehat{h_{\infty}}(\xi_{\infty})\widehat{h_f}(\xi_f)$,

$$\left(D^{\alpha,\beta}h\right)(x) = h_f(x_f)\left(D^{\beta}h_{\infty}\right)(x_{\infty}) + h_{\infty}(x_{\infty})\left(D^{\alpha}h_f\right)(x_f). \tag{4.39}$$

Note that $D^{\beta}h_{\infty} \in L^{2}(\mathbb{R}) \cap C(\mathbb{R},\mathbb{C})$, $D^{\alpha}h_f \in L^{2}(\mathbb{A}_f) \cap C(\mathbb{A}_f,\mathbb{C})$, cf. Lemma 92, and since $dx_{\mathbb{A}} = dx_{\infty}dx_{\mathbb{A}_f}$ we conclude $h_f D^{\beta}h_{\infty}$, $h_{\infty}D^{\alpha}h_f \in C(\mathbb{A},\mathbb{C}) \cap L^{2}(\mathbb{A})$. ∎

The space $\mathcal{D}(\mathbb{A})$ is not invariant under the action of the operator $D^{\alpha,\beta}$. To overcome such an inconvenience, we introduce an adelic version of the Lizorkin space of the second kind. First we recall that the real Lizorkin space of test functions, see e.g. [101, §2] or [102, §25], is defined by

$$\mathcal{L}_0(\mathbb{R}) = \left\{f_{\infty} \in \mathcal{D}(\mathbb{R}); \int_{\mathbb{R}} x_{\infty}^{n}f_{\infty}(x_{\infty})\,dx_{\infty} = 0, \text{ for } n \in \mathbb{N}\right\}.$$

The real Lizorkin space can be equipped with the topology of the space $\mathcal{D}(\mathbb{R})$, which makes $\mathcal{L}_0(\mathbb{R})$ a complete space. The real Lizorkin space is invariant with respect to D^{β}, is dense in $L^{p}(\mathbb{R})$, $1 < p < \infty$, and admits the following characterization: $f_{\infty} \in \mathcal{L}_0(\mathbb{R})$ if and only if $f_{\infty} \in \mathcal{D}(\mathbb{R})$ and

$$\frac{d^{n}}{d\xi_{\infty}^{n}}\mathcal{F}f(\xi_{\infty})\bigg|_{\xi_{\infty}=0} = 0, \qquad \text{for } n \in \mathbb{N}.$$

We introduce an *adelic Lizorkin space of the second kind* $\mathcal{L}_0 := \mathcal{L}_0(\mathbb{A})$ as

$$\mathcal{L}_0(\mathbb{A}) = \mathcal{L}_0(\mathbb{R}) \otimes \mathcal{L}_0(\mathbb{A}_f).$$

The space $\mathcal{L}_0(\mathbb{A})$ consists of finite linear combinations of factorizable functions $h(x) = h_\infty(x_\infty)h_f(x_f)$ with $h_\infty \in \mathcal{L}_0(\mathbb{R})$, $h_f \in \mathcal{L}_0(\mathbb{A}_f)$. Note that $\mathcal{L}_0(\mathbb{A})$ is a subspace of $\mathcal{D}(\mathbb{A})$ and it may be equipped with the topology of $\mathcal{D}(\mathbb{A})$.

Lemma 98 *With the above notation the following assertions hold:*

(i) $D^{\alpha,\beta}\mathcal{L}_0 = \mathcal{L}_0$ for $\alpha, \beta > 0$;
(ii) $\mathcal{L}_0$ is dense in $L^2(\mathbb{A})$.

Proof

(i) It is sufficient to consider a factorizable function $h = h_\infty h_f$, $h_\infty \in \mathcal{L}_0(\mathbb{R})$, $h_f \in \mathcal{L}_0(\mathbb{A}_f)$. Since $D^\alpha h_f \in \mathcal{L}_0(\mathbb{A}_f)$ and $D^\beta h_\infty \in \mathcal{L}_0(\mathbb{R})$, see Lemma 93 and [101, (9.1)], we conclude from (4.39) that $D^{\alpha,\beta}\mathcal{L}_0(\mathbb{A}) \subset \mathcal{L}_0(\mathbb{A})$.

Conversely, take $h \in \mathcal{L}_0(\mathbb{A})$. We want to show that the equation $D^{\alpha,\beta}g = h$ has a solution $g \in \mathcal{L}_0(\mathbb{A})$. We may assume without loss of generality that $h = h_\infty h_f$, $h_\infty \in \mathcal{L}_0(\mathbb{R})$, $h_f \in \mathcal{L}_0(\mathbb{A}_f)$. Applying the Fourier transform we obtain

$$\hat{g}(\xi) = \frac{\hat{h}_\infty(\xi_\infty)\hat{h}_f(\xi_f)}{|\xi_\infty|_\infty^\beta + \|\xi_f\|^\alpha}. \tag{4.40}$$

Since $h_f \in \mathcal{L}_0(\mathbb{A}_f)$, it follows from (4.32) that

$$\hat{h}_f(\xi_f) = \sum_{i=1}^{N} c_i 1_{B_R(\xi_i)}(\xi_f),$$

where the balls $B_R(\xi_i)$ are disjoint and $0 \notin B_R(\xi_i)$ for $i = 1, \ldots, N$. It follows from non-Archimedean property that the function $\|\xi_f\|$ is constant on each of the balls $B_R(\xi_i)$, hence we may rewrite (4.40) as

$$\hat{g}(\xi) = \sum_{i=1}^{N} \frac{c_i \hat{h}_\infty(\xi_\infty)}{|\xi_\infty|_\infty^\beta + d_i} 1_{B_R(\xi_i)}(\xi_f),$$

where $d_i := \|\xi_i\|^\alpha > 0$ are constants. It may be easily checked that the functions $\frac{c_i \hat{h}_\infty(\xi_\infty)}{|\xi_\infty|_\infty^\beta + d_i}$ are Fourier transforms of real Lizorkin functions and the functions $1_{B_R(\xi_i)}(\xi_f)$ are Fourier transforms of Lizorkin functions on $\mathbb{A}_f$, thus $g \in \mathcal{L}_0(\mathbb{A})$.

(ii) Since $\mathcal{L}_0(\mathbb{R})$ is dense in $L^2(\mathbb{R})$, see [101, Thm. 3.2], and $\mathcal{L}_0(\mathbb{A}_f)$ is dense in $L^2(\mathbb{A}_f)$ by Lemma 93, the tensor product $\mathcal{L}_0(\mathbb{A}) = \mathcal{L}_0(\mathbb{R}) \otimes \mathcal{L}_0(\mathbb{A}_f)$ is dense in the tensor product $L^2(\mathbb{R}) \otimes L^2(\mathbb{A}_f)$ which is isomorphic to the space $L^2(\mathbb{A})$, see e.g. [95, Theorem II.10]. ∎

Similarly to Sect. 4.3.4 we may consider the operator $\boldsymbol{D}^{\alpha,\beta}$ as an operator acting on $L^2(\mathbb{A})$. It is easy to see that the operator $\boldsymbol{D}^{\alpha,\beta}$ with the domain $Dom(\boldsymbol{D}^{\alpha,\beta}) = \mathcal{D}(\mathbb{A})$ is symmetric. Moreover, similarly to the proof of Lemma 98 (i) we may check that $(\boldsymbol{D}^{\alpha,\beta} \pm i)\mathcal{L}_0(\mathbb{A}) = \mathcal{L}_0(\mathbb{A})$, i.e. the ranges of the operators $\boldsymbol{D}^{\alpha,\beta} \pm i$ are dense in $L^2(\mathbb{A}_f)$, hence the operator $\boldsymbol{D}^{\alpha,\beta}$ is essentially self-adjoint. The following description of the closure holds.

Lemma 99 *The closure of the operator $\boldsymbol{D}^{\alpha,\beta}$, $\alpha, \beta > 0$ (let us denote it by $\boldsymbol{D}^{\alpha,\beta}$ again) with domain*

$$Dom\left(\boldsymbol{D}^{\alpha,\beta}\right) := \left\{ f \in L^2(\mathbb{A}) ; \left(|\xi_\infty|_\infty^\beta + \|\xi\|^\alpha \right) \hat{f} \in L^2(\mathbb{A}) \right\} \tag{4.41}$$

is a self-adjoint operator. Moreover, the following assertions hold:

(i) $\boldsymbol{D}^{\alpha,\beta} \geq 0$;
(ii) $\boldsymbol{D}^{\alpha,\beta}$ is m-accretive, i.e. $-\boldsymbol{D}^{\alpha,\beta}$ is an m-dissipative operator;
(iii) the spectrum $\sigma(\boldsymbol{D}^{\alpha,\beta}) = [0, \infty)$;
(iv) $-\boldsymbol{D}^{\alpha,\beta}$ is the infinitesimal generator of a contraction C_0 semigroup $\left(\mathcal{T}_{\alpha,\beta}(t) \right)_{t \geq 0}$. Moreover, the semigroup $\left(\mathcal{T}_{\alpha,\beta}(t) \right)_{t \geq 0}$ is bounded holomorphic (or analytic) with angle $\pi/2$.

Proof The proofs of (i), (ii) and (iv) are similar to the corresponding proofs in Lemma 94. We outline only the proof of (iii).

Suppose $\lambda > 0$ is given. We show that the equation $(\boldsymbol{D}^{\alpha,\beta} - \lambda)g = h$ cannot be solved for some $h \in L^2(\mathbb{A})$. We pick a factorizable h and by applying the Fourier transform we get

$$\hat{g}(\xi) = \frac{\hat{h}_\infty(\xi_\infty)\hat{h}_f(\xi_f)}{|\xi_\infty|_\infty^\beta + \|\xi_f\|^\alpha - \lambda}.$$

We pick $\hat{h}_f(\xi_f) = 1_{B_R(\xi_0)}(\xi)$, where $\|\xi_0\|^\alpha < \lambda$ and $0 \notin B_R(\xi_0)$, then $\|\xi_f\| = \|\xi_0\|$ for any $\xi_f \in B_R(\xi_0)$ and

$$\hat{g}(\xi) = \frac{\hat{h}_\infty(\xi_\infty)}{|\xi_\infty|_\infty^\beta + \|\xi_0\|^\alpha - \lambda} 1_{B_R(\xi_0)}(\xi).$$

If we take $h_\infty \in \mathcal{D}(\mathbb{R})$ such that $\hat{h}_\infty\left((\lambda - \|\xi_0\|^\alpha)^{1/\beta} \right) \neq 0$, then the function $\frac{\hat{h}_\infty(\xi_\infty)}{|\xi_\infty|_\infty^\beta + \|\xi_0\|^\alpha - \lambda} \notin L^2(\mathbb{R})$, and a fortiori $\hat{g} \notin L^2(\mathbb{A})$, hence $\lambda \in \sigma(\boldsymbol{D}^{\alpha,\beta})$. ∎

The representation of the generated semigroup $\left(\mathcal{T}_{\alpha,\beta}(t) \right)_{t \geq 0}$, in the case $\alpha > 1$, $0 < \beta \leq 2$, is presented in detail in Theorem 127.

4.5 The Adelic Heat Kernel on $\mathbb{A}_f$

In this section we introduce the adelic heat kernel on $\mathbb{A}_f$ as the inverse Fourier transform of $e^{-t\|y\|^\alpha}$ with $y \in \mathbb{A}_f$, $\|y\|$ defined by (4.5), $\alpha > 1$ and $t > 0$.

In Sects. 4.5, 4.6 and 4.7 we work only with finite adeles, for this reason in the variables we omit the subindex 'f'.

Proposition 100 *Consider the function* $\|y\|^\beta e^{-t\|y\|^\alpha}$ *for fixed* $t > 0$, $\beta \geq 0$ *and* $\alpha > 1$. *Then*

$$\|y\|^\beta e^{-t\|y\|^\alpha} \in L^\varrho \left(\mathbb{A}_f, dy_{\mathbb{A}_f}\right)$$

for any $1 \leq \varrho < +\infty$.

Proof It is sufficient to show that for any $t > 0$ and $\beta \geq 0$

$$I(t) := \int_{\mathbb{A}_f} \|y\|^\beta e^{-t\|y\|^\alpha} dy_{\mathbb{A}_f} < +\infty.$$

According to Lemmas 78 and 76

$$\int_{\mathbb{A}_f} \|y\|^\beta e^{-t\|y\|^\alpha} dy_{\mathbb{A}_f} = \sum_{p^m,\, m \neq 0} p^{m\beta} e^{-tp^{m\alpha}} \left(\Phi(p^m) - \Phi(p_{-}^m)\right),$$

thus we have to prove the convergence of the latter series. We consider two cases: $m < 0$ and $m > 0$.

If $m < 0$, then $p^{m\beta} e^{-tp^{m\alpha}} \leq 1$ and

$$S_{-}(t) := \sum_{p^m,\, m<0} p^{m\beta} e^{-tp^{m\alpha}} \left(\Phi(p^m) - \Phi(p_{-}^m)\right)$$

$$\leq \sum_{p^m,\, m<0} \left(\Phi(p^m) - \Phi(p_{-}^m)\right) = \Phi(1/2).$$

Let $m > 0$. We have

$$S_{+}(t) := \sum_{p^m,\, m>0} p^{m\beta} e^{-tp^{m\alpha}} \left(\Phi(p^m) - \Phi(p_{-}^m)\right) \leq \sum_{p^m,\, m>0} p^{m\beta} e^{-tp^{m\alpha}} \Phi(p^m).$$

We recall that the Prime Number Theorem is equivalent to

$$\ln \Phi(x) = \psi(x) \sim x, \quad x \to \infty,$$

see e.g. [33], hence there exists a constant C such that

$$S_+(t) \leq \sum_p \sum_{m=1}^{\infty} p^{m\beta} e^{-tp^{m\alpha} + Cp^m}.$$

We want to show the existence of a positive constant $M = M(\beta)$ such that

$$p^{m\beta} e^{-tp^{m\alpha} + Cp^m} \leq M p^{-1-m} \quad \text{for all } m \geq 1 \text{ and prime } p,$$

or equivalently that $p^{1+m(\beta+1)} e^{-tp^{m\alpha} + Cp^m} \leq M$. Since $p^{1+m(\beta+1)} \leq e^{(\beta+1)p^m}$ for all $m \geq 1$ and $p \geq 2$, consider $e^{(C+\beta+1)p^m - tp^{\alpha m}}$. This expression is less than or equal to 1 when $p^m \geq \left(\frac{C+\beta+1}{t}\right)^{\frac{1}{\alpha-1}}$ and hence there exist only a finite number of pairs (p, m) for which it can be greater than 1, so the announced constant exists. Therefore

$$S_+(t) \leq M \sum_p \sum_{m=1}^{\infty} p^{-1-m} \leq 2M \sum_p p^{-2} < +\infty. \qquad \blacksquare$$

Definition 101 We define *the adelic heat kernel* on $\mathbb{A}_f$ as

$$Z(x, t; \alpha) := Z(x, t) = \int_{\mathbb{A}_f} \chi(-\xi x) e^{-t\|\xi\|^{\alpha}} d\xi_{\mathbb{A}_f}, \quad x \in \mathbb{A}_f, \ t > 0, \ \alpha > 1.$$
$$(4.42)$$

By Proposition 100 the integral is convergent. When considering $Z(x, t)$ as a function of x for t fixed we will write $Z_t(x)$. By applying Theorem 79 to the function $e^{-t\|\xi\|^{\alpha}}$ we obtain the following result.

Proposition 102 *The following representation holds for the heat kernel:*

$$Z(x, t) = \sum_{q^j < \|x\|^{-1}, \, j \neq 0} \Phi(q^j) \left(e^{-tq^{j\alpha}} - e^{-t(q_+^j)^{\alpha}} \right) \quad \text{for } t > 0, \ x \in \mathbb{A}_f, \qquad (4.43)$$

where q^j runs through all non-zero powers of prime numbers; functions $\|x\|$, $\Phi(x)$ and q_+^j are defined by (4.5), (4.11) and (4.12). For $x = 0$ the expression $\|0\|^{-1}$ in the representation means ∞.

Lemma 103 *The following estimate holds for the heat kernel:*

$$Z(x, t) \leq 2t\|x\|^{-\alpha} \Phi(\|x\|_-^{-1}), \quad x \in \mathbb{A}_f \setminus \{0\}, \ t > 0. \qquad (4.44)$$

Proof From the inequality $1 - e^{-x} \leq x$ valid for $x \geq 0$ we obtain

$$e^{-tq^{j\alpha}} - e^{-t(q_+^j)^{\alpha}} \leq 1 - e^{-t(q_+^j)^{\alpha}} \leq t(q_+^j)^{\alpha}.$$

Then with the use of the inequality $\frac{1}{2}\Phi(q^j) \leq \Phi(q^j) - \Phi(q_-^j)$ we have

$$Z(x,t) = \sum_{q^j < \|x\|^{-1}} \Phi(q^j)\left(e^{-tq^{j\alpha}} - e^{-t(q_+^j)^\alpha}\right) \leq t \sum_{q^j < \|x\|^{-1}} \Phi(q^j)(q_+^j)^\alpha$$

$$\leq 2t\|x\|^{-\alpha} \sum_{q^j < \|x\|^{-1}} \left(\Phi(q^j) - \Phi(q_-^j)\right) = 2t\|x\|^{-\alpha}\Phi(\|x\|_-^{-1}). \qquad \blacksquare$$

Corollary 104 *With the above notation the following assertions hold:*

(i) $Z(x,t) \geq 0$ *for* $t > 0$;
(ii) $\lim_{t\to 0+} Z(x,t) = 0$ *for any* $x \in \mathbb{A}_f \setminus \{0\}$;
(iii) *For any* $\epsilon > 0$ *there exists a constant* $C = C(\epsilon)$ *such that for any* $t > 0$

$$\int_{\|y\|>\epsilon} Z_t(y)\, dy_{\mathbb{A}_f} \leq Ct < +\infty. \qquad (4.45)$$

Proof The statements (i) and (ii) immediately follows from formulas (4.43) and (4.44), respectively.

(iii) By (4.44) and Lemma 78 we have

$$\int_{\|y\|>\epsilon} Z_t(y)\, dy_{\mathbb{A}_f} \leq \sum_{p^k>\epsilon} \Phi(p^k) \cdot 2tp^{-k\alpha}\Phi(p^{-k}) = 2t \cdot \sum_{p^k>\epsilon} p^{-k\alpha} < +\infty,$$

where we have used (4.14) and (4.15). $\qquad \blacksquare$

Theorem 105 *The adelic heat kernel on* $\mathbb{A}_f$ *satisfies the following:*

(i) $Z(x,t) \geq 0$ *for any* $t > 0$;
(ii) $\int_{\mathbb{A}_f} Z_t(x)\, dx_{\mathbb{A}_f} = 1$ *for any* $t > 0$;
(iii) $Z_t(x) \in L^1(\mathbb{A}_f)$ *for any* $t > 0$;
(iv) $Z_t(x) * Z_{t'}(x) = Z_{t+t'}(x)$ *for any* $t, t' > 0$;
(v) $\lim_{t\to 0+} Z_t(x) = \delta(x)$ *in* $\mathcal{D}'(\mathbb{A}_f)$;
(vi) $Z_t(x)$ *is a uniformly continuous function for any fixed* $t > 0$;
(vii) $Z(x,t)$ *is uniformly continuous in* t, *i.e.* $Z(x,t) \in C((0,\infty), C(\mathbb{A}_f))$ *or* $\lim_{t'\to t} \max_{x\in\mathbb{A}_f} |Z(x,t) - Z(x,t')| = 0$ *for any* $t > 0$.

Proof

(i) It follows from Corollary 104.
(ii) For any $t > 0$ the function $e^{-t\|\xi\|^\alpha}$ is continuous at $\xi = 0$ and by Proposition 100 we have $e^{-t\|\xi\|^\alpha} \in L^1(\mathbb{A}_f) \cap L^2(\mathbb{A}_f)$. Then $Z_t(x) \in C(\mathbb{A}_f, \mathbb{R}) \cap L^2(\mathbb{A}_f)$. Now the statement follows from the inversion formula for the Fourier transform on $\mathbb{A}_f$.
(iii) The statement follows from (i) and (ii).

(iv) By the previous property $Z_t(x) \in L^1(\mathbb{A}_f)$ for any $t > 0$. Then

$$Z_t(x) * Z_{t'}(x) = \mathcal{F}_{\xi \to x}^{-1}\left(e^{-t\|\xi\|^\alpha} e^{-t'\|\xi\|^\alpha}\right) = \mathcal{F}_{\xi \to x}^{-1}\left(e^{-(t+t')\|\xi\|^\alpha}\right) = Z_{t+t'}(x).$$

(v) Since $e^{-t\|\xi\|^\alpha} \in C(\mathbb{A}_f, \mathbb{R}) \cap L^1(\mathbb{A}_f)$, cf. Proposition 100, the scalar product

$$\left(e^{-t\|\xi\|^\alpha}, f(\xi)\right) = \int_{\mathbb{A}_f} e^{-t\|\xi\|^\alpha} f(\xi)\, d\xi_{\mathbb{A}_f} \quad \text{with } f \in \mathcal{D}(\mathbb{A}_f)$$

defines a distribution on $\mathbb{A}_f$. Since the support of f is compact, cf. Lemma 87, and $e^{-t\|\xi\|^\alpha} \in L^1(\mathbb{A}_f)$, the dominated convergence theorem with the characteristic function of the support of f as a dominant function implies

$$\lim_{t \to 0+}\left(e^{-t\|\xi\|^\alpha}, f(\xi)\right) = (1, f)$$

and then, as $\mathcal{F}(\mathcal{D}(\mathbb{A}_f)) = \mathcal{D}(\mathbb{A}_f)$, we have

$$\lim_{t \to 0+}(Z(x,t), f) = \lim_{t \to 0+}\left(\mathcal{F}_{\xi \to x}^{-1}(e^{-t\|\xi\|^\alpha}), f(x)\right) = \left(1, \mathcal{F}_{\xi \to x}^{-1} f\right) = (\delta, f).$$

(vi) Since $Z_t(x) = \mathcal{F}_{\xi \to x}^{-1}\left(e^{-t\|\xi\|^\alpha}\right)$ and $e^{-t\|\xi\|^\alpha} \in L^1(\mathbb{A}_f)$ for $t > 0$, $Z_t(x)$ is uniformly continuous in x for any fixed $t > 0$.

(vii) Suppose that $t < t'$. By the mean value theorem $e^{-t\|\xi\|^\alpha} - e^{-t'\|\xi\|^\alpha} = (t' - t)\|\xi\|^\alpha e^{-t(\|\xi\|)\|\xi\|^\alpha}$, where $t < t(\|\xi\|) < t'$. Hence

$$|Z(x,t) - Z(x,t')| = \left|\int_{\mathbb{A}_f} \chi(\xi \cdot x)\left(e^{-t\|\xi\|^\alpha} - e^{-t'\|\xi\|^\alpha}\right) d\xi_{\mathbb{A}_f}\right|$$

$$= |t - t'|\left|\int_{\mathbb{A}_f} \chi(\xi \cdot x)\|\xi\|^\alpha e^{-t(\|\xi\|)\|\xi\|^\alpha}\, d\xi_{\mathbb{A}_f}\right|$$

$$\leq |t - t'|\int_{\mathbb{A}_f} \|\xi\|^\alpha e^{-t_0\|\xi\|^\alpha}\, d\xi_{\mathbb{A}_f},$$

for some $0 < t_0 < t, t'$. Now the statement follows from Proposition 100. ∎

4.6 Markov Processes on $\mathbb{A}_f$

Along this section we consider $(\mathbb{A}_f, \rho)$ as the complete non-Archimedean metric space and use the terminology, notation and results of [39, Chapters Two, Three]. Let $\mathcal{B}$ denote the σ-algebra of the Borel sets of $\mathbb{A}_f$. Then $(\mathbb{A}_f, \mathcal{B}, dx_{\mathbb{A}_f})$ is a measure space. Let $1_B(x)$ denote the characteristic function of a set $B \in \mathcal{B}$.

We assume along this section that $\alpha > 1$ and set

$$p(t, x, y) := Z(x - y, t) \qquad \text{for } t > 0, \ x, y \in \mathbb{A}_f,$$

and

$$P(t, x, B) := \begin{cases} \int_B p(t, x, y) \, dy_{\mathbb{A}_f} & \text{for } t > 0, \ x \in \mathbb{A}_f, B \in \mathcal{B} \\ 1_B(x) & \text{for } t = 0. \end{cases}$$

Lemma 106 *With the above notation the following assertions hold:*

(i) $p(t, x, y)$ is a normal transition density;
(ii) $P(t, x, B)$ is a normal transition function.

Proof The result follows from Theorem 105, see [39, Sec. 2.1] for further details. ∎

Lemma 107 *The transition function $P(t, y, B)$ satisfies the following two conditions:*

(i) for each $u \geq 0$ and a compact B

$$\lim_{x \to \infty} \sup_{t \leq u} P(t, x, B) = 0; \qquad [\text{Condition } L(B)]$$

(ii) for each $\epsilon > 0$ and a compact B

$$\lim_{t \to 0+} \sup_{x \in B} P\big(t, x, \mathbb{A}_f \setminus B_\epsilon(x)\big) = 0. \qquad [\text{Condition } M(B)]$$

Proof Since B is a compact, $\text{dist}(x, B) =: d(x) \to \infty$ as $x \to \infty$. Since the function $\Phi(x)$ is non-decreasing, we obtain from (4.44) that $Z(x - y, t) \leq 2u\big(d(x)\big)^{-\alpha} \Phi\big((d(x))^{-1}\big)$ for any $y \in B$ and $t \leq u$. Hence $P(t, x, B) \leq 2u\big(d(x)\big)^{-\alpha} \cdot \Phi\big((d(x))^{-1}\big) \cdot \text{vol}(B) \to 0$ as $x \to \infty$.

To verify Condition $M(B)$ we proceed as follows: for $y \in \mathbb{A}_f \setminus B_\epsilon(x)$ we have $\|x - y\| > \epsilon$. The statement follows from (4.45):

$$P\big(t, x, \mathbb{A}_f \setminus B_\epsilon(x)\big) \leq C(\epsilon)t \to 0, \qquad t \to 0+.$$

∎

Theorem 108 *$Z(x, t)$ is the transition density of a time- and space homogenous Markov process which is bounded, right-continuous and has no discontinuities other than jumps.*

Proof The result follows from [39, Theorem 3.6], Remark 73 (ii) and Lemmas 106, 107. ∎

Remark 109 The more strict version of Condition $M(B)$ which is sufficient for the continuity of a Markov process, namely, that for each $\epsilon > 0$ and a compact B

$$\lim_{t \to 0+} \frac{1}{t} \sup_{x \in B} P\left(t, x, \mathbb{A}_f \setminus B_\epsilon(x)\right) = 0, \qquad [\text{Condition } N(B)]$$

does not hold for the function $Z(x, t)$. This may be easily seen if we take $\epsilon = 1/4$. In such case by Proposition 102 and Lemma 78 we have

$$\int_{\mathbb{A}_f \setminus B_{1/4}(x)} Z_t(x-y)\, dy_{\mathbb{A}_f} \geq \int_{S_{1/3}(x)} Z_t(x-y)\, dy_{\mathbb{A}_f}$$

$$= \text{vol } S_{1/3}(x) \cdot \sum_{q^j < 3, j \neq 0} \Phi\left(q^j\right) \left(e^{-tq^{j\alpha}} - e^{-t\left(q_+^j\right)^\alpha}\right)$$

$$\geq \frac{1}{3}\left(e^{-2^\alpha t} - e^{-3^\alpha t}\right),$$

hence

$$\lim_{t \to 0+} \frac{1}{t} \sup_{x \in B} P\left(t, x, \mathbb{A}_f \setminus B_{1/4}(x)\right) \geq \frac{3^\alpha - 2^\alpha}{3} \neq 0.$$

4.7 Cauchy Problem for Parabolic Type Equations on $\mathbb{A}_f$

Consider the following Cauchy problem

$$\begin{cases} \dfrac{\partial u(x, t)}{\partial t} + \mathbf{D}^\alpha u(x, t) = 0, & x \in \mathbb{A}_f, \ t \in [0, +\infty) \\[2mm] u(x, 0) = u_0(x), & u_0(x) \in Dom(\mathbf{D}^\alpha), \end{cases} \tag{4.46}$$

where $\alpha > 1$, $\mathbf{D}^\alpha$ is the pseudodifferential operator defined by (4.33) with the domain given by (4.35) and $u : \mathbb{A}_f \times [0, \infty) \to \mathbb{C}$ is an unknown function.

We say that a function $u(x, t)$ is a *solution* of (4.46) if $u \in C([0, \infty), Dom(\mathbf{D}^\alpha)) \cap C^1([0, \infty), L^2(\mathbb{A}_f))$ and u satisfies equation (4.46) for all $t \geq 0$.

We understand the notions of continuity in t, differentiability in t and equalities in the $L^2(\mathbb{A}_f)$ sense, as it is customary in the semigroup theory. More precisely, we say that a function $u(x, t)$ is continuous in t at t_0 if $\lim_{t \to t_0} \|u(x, t) - u(x, t_0)\|_{L^2(\mathbb{A}_f)} = 0$; the function $u'_t(x, t)$ is the time derivative of function $u(x, t)$ at t_0 if $\lim_{t \to t_0} \left\| \frac{u(x,t) - u(x,t_0)}{t - t_0} - u'_t(x, t_0) \right\|_{L^2(\mathbb{A}_f)} = 0$; two functions $f(x, t)$ and $g(x, t)$ are equal at t_0 if $\|f(x, t_0) - g(x, t_0)\|_{L^2(\mathbb{A}_f)} = 0$.

We know from Lemma 94 that the operator $-\mathbf{D}^\alpha$ generates a C_0 semigroup. Therefore Cauchy problem (4.46) is well-posed, i.e. it is uniquely solvable with the solution continuously dependent on the initial data, and its solution is given by $u(x, t) = T(t)u_0(x)$, $t \geq 0$, see e.g. [7, 24, 41]. However the general theory does not give an explicit formula for the semigroup $(T(t))_{t \geq 0}$. We show that the operator $T(t)$ for $t > 0$ coincides with the operator of convolution with the heat kernel $Z_t * \cdot$. In order to prove this, we first construct a solution of Cauchy problem 4.46 with the initial value from $\mathcal{D}(\mathbb{A}_f)$ without using the semigroup theory. Then we extend the result to all initial values from $Dom(\boldsymbol{D}^\alpha)$, see Proposition 112 and Theorem 114.

We show in Theorem 117 that in the case $u_0 \in \mathcal{L}_0(\mathbb{A}_f)$, the function $u(x, t)$ is the solution of Cauchy problem 4.46 in a stricter sense, i.e. $u(x, t) \in C^1([0, \infty), \mathcal{L}_0(\mathbb{A}_f))$ and all limits and equalities are understood pointwise.

4.7.1 Homogeneous Equations with Initial Values in $\mathcal{D}(\mathbb{A}_f)$

We first consider Cauchy problem 4.46 with the initial value from the space $\mathcal{D}(\mathbb{A}_f)$. To simplify notations, set $Z_0 * u_0 = (Z_t * u_0)|_{t=0} := u_0$. Note that such definition is consistent with Theorem 105 (v). We define the function

$$u(x, t) = Z_t(x) * u_0(x), \qquad t \geq 0. \tag{4.47}$$

Since $Z_t(x) \in L^1(\mathbb{A}_f)$ for $t > 0$ and $u_0(x) \in \mathcal{D}(\mathbb{A}_f) \subset L^\infty(\mathbb{A}_f)$, the convolution exists and is a continuous function, see [100, Theorem 1.1.6].

Lemma 110 *Let $u_0 \in \mathcal{D}(\mathbb{A}_f)$ and $u(x, t)$, $t \geq 0$ is defined by (4.47). Then $u(x, t)$ is continuously differentiable in time for $t \geq 0$ and the derivative is given by*

$$\frac{\partial u}{\partial t}(x, t) = -\mathcal{F}_{\xi \to x}^{-1}\left(\|\xi\|^\alpha e^{-t\|\xi\|^\alpha} \cdot 1_{B_R}(\xi)\right) * u_0(x), \tag{4.48}$$

where $1_{B_R}(\cdot)$ is the characteristic function of the ball B_R, $R = (1/\ell)_-$ and ℓ is the parameter of constancy of the function u_0, see (4.31) and Proposition 89.

Proof Let $h_t(x)$ be a function defined by the right-hand side of (4.48). Since $\|\xi\|^\alpha e^{-t\|\xi\|^\alpha} \cdot 1_{B_R}(\xi) \in L^1(\mathbb{A}_f) \cap L^2(\mathbb{A}_f)$ for any $t \geq 0$, the function $h_t(x)$ is well-defined and belongs to $C(\mathbb{A}_f) \cap L^2(\mathbb{A}_f)$.

Let $t_0 \geq 0$. Consider a limit

$$\lim_{t \to t_0} \left\| \frac{u(x, t) - u(x, t_0)}{t - t_0} - h_t(x, t_0) \right\|_{L^2(\mathbb{A}_f)}$$

$$= \lim_{t \to t_0} \left\| \frac{e^{-t\|\xi\|^\alpha} - e^{-t_0\|\xi\|^\alpha}}{t - t_0} \hat{u}_0(\xi) + \|\xi\|^\alpha e^{-t_0\|\xi\|^\alpha} 1_{B_R}(\xi) \cdot \hat{u}_0(\xi) \right\|_{L^2(\mathbb{A}_f)}$$

$$= \lim_{t \to t_0} \left\| \left(\frac{e^{-t\|\xi\|^\alpha} - e^{-t_0\|\xi\|^\alpha}}{t - t_0} + \|\xi\|^\alpha e^{-t_0\|\xi\|^\alpha} \right) 1_{B_R}(\xi) \cdot \hat{u}_0(\xi) \right\|_{L^2(\mathbb{A}_f)},$$

where we have applied Steklov-Parseval formula and the fact that $\operatorname{supp} \hat{u}_0 \subset B_R$ which follows from Proposition 89. By applying the mean value theorem twice we obtain

$$\frac{e^{-t\|\xi\|^\alpha} - e^{-t_0\|\xi\|^\alpha}}{t - t_0} + \|\xi\|^\alpha e^{-t_0\|\xi\|^\alpha} = -\|\xi\|^\alpha e^{-t'\|\xi\|^\alpha} + \|\xi\|^\alpha e^{-t_0\|\xi\|^\alpha}$$

$$= (t' - t_0)\|\xi\|^{2\alpha} e^{-t''\|\xi\|^\alpha},$$

where $t' = t'(\|\xi\|)$ is a point between t_0 and t and $t'' = t''(\|\xi\|)$ is a point between t_0 and t' (and thus between t_0 and t). Hence

$$\left\| \left(\frac{e^{-t\|\xi\|^\alpha} - e^{-t_0\|\xi\|^\alpha}}{t - t_0} + \|\xi\|^\alpha e^{-t_0\|\xi\|^\alpha} \right) 1_{B_R}(\xi) \cdot \hat{u}_0(\xi) \right\|_{L^2(\mathbb{A}_f)}$$

$$\leq |t - t_0| R^{2\alpha} \|\hat{u}_0(\xi)\|_{L^2(\mathbb{A}_f)} \to 0, \qquad t \to t_0,$$

i.e. $h_t(x)$ is the time derivative of the function $u(x,t)$ for any $t \geq 0$.

The proof of the continuous differentiability in time of $u(x,t)$ follows from the time continuity of $h_t(x)$ which can be checked similarly. ∎

Lemma 111 *Let $u_0 \in \mathcal{D}(\mathbb{A}_f)$ and $u(x,t)$, $t \geq 0$ is defined by (4.47). Then $u(x,t) \in \operatorname{Dom}(\boldsymbol{D}^\alpha)$ for any $t \geq 0$ and*

$$\boldsymbol{D}^\alpha u(x,t) = \mathcal{F}^{-1}_{\xi \to x}\left(\|\xi\|^\alpha e^{-t\|\xi\|^\alpha} \cdot 1_{B_R}(\xi) \right) * u_0(x), \qquad (4.49)$$

where $1_{B_R}(\cdot)$ is the characteristic function of the ball B_R, $R = (1/\ell)_-$ and ℓ is the parameter of constancy of the function u_0, see (4.31) and Proposition 89.

Proof Note that $\hat{u}_0 \in \mathcal{D}(\mathbb{A}_f)$ which implies that $e^{-t\|\xi\|^\alpha} \cdot \hat{u}_0(\xi) \in L^1(\mathbb{A}_f) \cap L^2(\mathbb{A}_f)$ and $\|\xi\|^\alpha e^{-t\|\xi\|^\alpha} \cdot \hat{u}_0(\xi) \in L^1(\mathbb{A}_f) \cap L^2(\mathbb{A}_f)$ for any $t \geq 0$. Hence we may calculate $\boldsymbol{D}^\alpha u(x,t)$ by formula (4.33). For $t > 0$ we obtain

$$\boldsymbol{D}^\alpha u(x,t) = \mathcal{F}^{-1}_{\xi \to x}\left(\|\xi\|^\alpha \cdot \hat{u}(\xi,t) \right) = \mathcal{F}^{-1}_{\xi \to x}\left(\|\xi\|^\alpha \widehat{Z_t}(\xi) \cdot \hat{u}_0(\xi) \right)$$

$$= \mathcal{F}^{-1}_{\xi \to x}\left(\|\xi\|^\alpha e^{-t\|\xi\|^\alpha} 1_{B_R}(\xi) \cdot \hat{u}_0(\xi) \right) = \mathcal{F}^{-1}_{\xi \to x}\left(\|\xi\|^\alpha e^{-t\|\xi\|^\alpha} \cdot 1_{B_R}(\xi) \right) * u_0(x),$$

where we have used the fact that $\operatorname{supp} \hat{u}_0 \subset B_R$.

For $t = 0$ we obtain

$$\boldsymbol{D}^\alpha u(x,0) = \boldsymbol{D}^\alpha u_0(x) = \mathcal{F}^{-1}_{\xi \to x}\left(\|\xi\|^\alpha \cdot \hat{u}_0(\xi) \right)$$

$$= \mathcal{F}^{-1}_{\xi \to x}\left(\|\xi\|^\alpha e^{-0\|\xi\|^\alpha} 1_{B_R}(\xi) \cdot \hat{u}_0(\xi) \right) = \mathcal{F}^{-1}_{\xi \to x}\left(\|\xi\|^\alpha e^{-0\|\xi\|^\alpha} \cdot 1_{B_R}(\xi) \right) * u_0(x).$$

∎

As an immediate consequence from Lemmas 110 and 111 we obtain:

Proposition 112 *Let the function $u_0 \in \mathcal{D}(\mathbb{A}_f)$. Then the function $u(x, t)$ defined by (4.47) is a solution of Cauchy problem (4.46).*

4.7.2 Homogeneous Equations with Initial Values in $L^2(\mathbb{A}_f)$

Consider the operator $T(t)$, $t \geq 0$ of convolution with the heat kernel, i.e.

$$T(t)u = Z_t * u. \tag{4.50}$$

Since $Z_t \in L^2(\mathbb{A}_f)$, the convolution $Z_t * u$ is a continuous function of x for $t > 0$ and any $u \in L^2(\mathbb{A}_f)$, see [100, Theorem 1.1.6].

Lemma 113 *The operator $T(t) : L^2(\mathbb{A}_f) \to L^2(\mathbb{A}_f)$ is bounded.*

Proof Consider a function $u \in L^2(\mathbb{A}_f)$. Since $Z_t \in L^1(\mathbb{A}_f)$, see Theorem 105 (iii), by the Young inequality and Theorem 105 (ii)

$$\|Z_t * u\|_{L^2} \leq \|Z_t\|_{L^1} \cdot \|u\|_{L^2} = \|u\|_{L^2}.$$

Hence $T(t)u = Z_t * u \in L^2(\mathbb{A}_f)$ and $\|T(t)\| \leq 1$. ∎

Theorem 114 *Let $\alpha > 1$. Then the following assertions hold.*

(i) *The operator $-D^\alpha$ generates a C_0 semigroup $\left(\mathcal{T}(t)\right)_{t \geq 0}$. The operator $\mathcal{T}(t)$ coincides for each $t \geq 0$ with the operator $T(t)$ given by (4.50).*

(ii) *Cauchy problem 4.46 is well-posed and its solution is given by $u(x, t) = Z_t * u_0$, $t \geq 0$.*

Proof According to Lemma 94, the operator $-D^\alpha$ generates a C_0 semigroup $\left(\mathcal{T}(t)\right)_{t \geq 0}$. Hence Cauchy problem 4.46 is well-posed, see e.g. [24, Theorem 3.1.1]. By Proposition 112, $\mathcal{T}(t)|_{\mathcal{D}(\mathbb{A}_f)} = T(t)|_{\mathcal{D}(\mathbb{A}_f)}$ and both operators $\mathcal{T}(t)$ and $T(t)$ are defined on the whole $L^2(\mathbb{A}_f)$ and bounded. By the continuity we conclude that $\mathcal{T}(t) = T(t)$ on $L^2(\mathbb{A}_f)$. Now the statements follow from well-known results of the semigroup theory, see e.g. [7, Proposition 3.1.9.], [24, Theorem 3.1.1], [41, Ch. 2, Proposition 6.2]. ∎

Remark 115 Since the semigroup $\left(\mathcal{T}(t)\right)_{t \geq 0}$ is holomorphic, Cauchy problem 4.46 possesses *smoothing effect*, see e.g. [7, Corollary 3.7.21]. More precisely, consider Cauchy problem 4.46 with weaker requirement on the initial value, namely, let $u_0 \in L^2(\mathbb{A}_f)$. Then there exists a unique function

$$u(x, t) \in C\big([0, \infty), L^2(\mathbb{A}_f)\big) \cap C\big((0, \infty), Dom(D^\alpha)\big) \cap C^\infty\big((0, \infty), L^2(\mathbb{A}_f)\big)$$

satisfying the equation for $t > 0$ and satisfying the initial condition. That is, this weaker Cauchy problem is solvable for arbitrary initial data and the solution is infinitely differentiable in t for $t > 0$.

4.7.3 Homogeneous Equations with Initial Values in $\mathcal{L}_0(\mathbb{A}_f)$

We now consider the Cauchy problem

$$
\begin{cases}
\dfrac{\partial u(x,t)}{\partial t} + \boldsymbol{D}^\alpha u(x,t) = 0, & x \in \mathbb{A}_f,\ t \in [0, +\infty) \\
u(x,0) = u_0(x), & u_0(x) \in \mathcal{L}_0(\mathbb{A}_f),
\end{cases}
\tag{4.51}
$$

with the initial value from the space $\mathcal{L}_0(\mathbb{A}_f)$ and the pseudodifferential operator $\boldsymbol{D}^\alpha$ with the smaller domain $Dom(\boldsymbol{D}^\alpha) = \mathcal{L}_0(\mathbb{A}_f)$.

We say that a function $u(x,t)$ is *a classical solution of* (4.51), if $u \in C^1([0,\infty), \mathcal{L}_0(\mathbb{A}_f))$ and u satisfies equation (4.51) for all $t \geq 0$, with the understanding that all the involved limits are taken in the topology of $\mathcal{L}_0(\mathbb{A}_f)$.

Lemma 116 *Let* $u_0 \in \mathcal{L}_0(\mathbb{A}_f)$ *and the function* $u(x,t)$ *is defined by (4.47). Then* $u(x,t) \in \mathcal{L}_0(\mathbb{A}_f)$ *for any* $t > 0$.

Proof Since $e^{-t\|\xi\|^\alpha}$ is locally constant outside of the origin, the function $h_t(\xi) = e^{-t\|\xi\|^\alpha} \cdot \hat{u}_0(\xi) \in \mathcal{D}(\mathbb{A}_f)$ with $h_t(0) = 0$. Then $\mathcal{F}^{-1}_{\xi \to x}\left(e^{-t\|\xi\|^\alpha} \cdot \hat{u}_0(\xi)\right) = Z_t(x) * u_0(x) \in \mathcal{L}_0(\mathbb{A}_f)$ for $t > 0$. ∎

Theorem 117 *Let the function* $u_0 \in \mathcal{L}_0(\mathbb{A}_f)$. *Then the function* $u(x,t)$ *defined by (4.47) is the classical solution of Cauchy problem 4.51.*

Proof By Lemma 116 the function $u(x,t)$ is correctly defined and $u(x,t) \in Dom(\boldsymbol{D}^\alpha) = \mathcal{L}_0(\mathbb{A}_f)$ for all $t \geq 0$.

We assert that there exist constants ℓ and R not dependent on t such that $u(x,t) \in \mathcal{D}^\ell_R$ and $\boldsymbol{D}^\alpha u(x,t) \in \mathcal{D}^\ell_R$ for all $t \geq 0$. Consider the function $h_t(\xi) = e^{-t\|\xi\|^\alpha} \cdot \hat{u}_0(\xi)$, $t \geq 0$. Since $u_0 \in \mathcal{L}_0$, the function $e^{-t\|\xi\|^\alpha}$ is locally constant on the support of $\hat{u}_0$. Moreover, the parameter of constancy of $e^{-t\|\xi\|^\alpha}$ on the support of $\hat{u}_0$ does not depend on t. Hence there exist parameters ℓ' and R' such that $h_t \in \mathcal{D}^{\ell'}_{R'}$ for any $t \geq 0$. By Proposition 89 we have $\left(\mathcal{F}^{-1}h_t\right)(x) = u(x,t) \in \mathcal{D}^{(1/R')-}_{(1/\ell')-}$ for any $t \geq 0$. Similar proof works for the function $g_t(\xi) := \mathcal{F}(\boldsymbol{D}^\alpha u(x,t)) = \|\xi\|^\alpha e^{-t\|\xi\|^\alpha} \cdot \hat{u}_0(\xi)$.

We recall that for finite dimensional spaces, the uniform convergence is equivalent to the L^2-convergence. Since $\mathcal{D}^\ell_R$ is a finite dimensional space, cf. Proposition 89, by applying Lemmas 110 and 111, we have $\frac{\partial u}{\partial t}(x,t) \in \mathcal{L}_0(\mathbb{A}_f)$, $u(x,t)$ is a solution of Cauchy problem (4.51) and u belongs to $C^1([0,\infty), \mathcal{L}_0(\mathbb{A}_f))$. ∎

4.7.4 Non Homogeneous Equations

Consider the following Cauchy problem

$$
\begin{cases}
\dfrac{\partial u(x,t)}{\partial t} + D^{\alpha} u(x,t) = f(x,t), & x \in \mathbb{A}_f, \ t \in [0,T], \ T > 0 \\[2mm]
u(x,0) = u_0(x), & u_0(x) \in Dom(D^{\alpha}).
\end{cases}
\tag{4.52}
$$

We say that a function $u(x,t)$ is a *solution* of (4.52), if u belongs to $C([0,T], Dom(D^{\alpha})) \cap C^1([0,T], L^2(\mathbb{A}_f))$ and if u satisfies equation (4.52) for $t \in [0,T]$.

Theorem 118 *Let $\alpha > 1$ and let $f \in C([0,T], L^2(\mathbb{A}_f))$. Assume that at least one of the following conditions is satisfied:*

(i) $f \in L^1((0,T), Dom(D^{\alpha}))$;
(ii) $f \in W^{1,1}((0,T), L^2(\mathbb{A}_f))$.

   Then Cauchy problem (4.52) has a unique solution given by

$$
u(x,t) = \int_{\mathbb{A}_f} Z(x-y,t) u_0(y) \, dy_{\mathbb{A}_f}
$$

$$
+ \int_0^t \left\{ \int_{\mathbb{A}_f} Z(x-y, t-\tau) f(y,\tau) \, dy_{\mathbb{A}_f} \right\} d\tau.
$$

Proof With the use of Theorem 114 the proof follows from well-known results of the semigroup theory, see e.g. [7, Proposition 3.1.16], [24, Proposition 4.1.6]. ∎

4.8 The Adelic Heat Kernel on $\mathbb{A}$

We recall that the *Archimedean heat kernel* is defined as

$$
Z(x_\infty, t; \beta) = \int_{\mathbb{R}} \chi_\infty(\xi_\infty x_\infty) e^{-t|\xi_\infty|_\infty^{\beta}} d\xi_\infty, \quad t > 0, \ \beta \in (0,2].
$$

This heat kernel is a solution of the pseudodifferential equation

$$
\frac{\partial u(x_\infty, t)}{\partial t} + \mathcal{F}^{-1}_{\xi_\infty \to x_\infty} \left(|\xi_\infty|_\infty^{\beta} \, \mathcal{F}^{-1}_{x_\infty \to \xi_\infty} u(x_\infty, t) \right) = 0.
$$

For a more detailed discussion of the Archimedean heat kernel and its properties the reader may consult [38, Section 2] and references therein.

   From now on we will denote heat kernel (4.42) as $Z(x_f, t; \alpha)$.

Definition 119 For fixed $\alpha > 1$, $\beta \in (0, 2]$ we define *the heat kernel on* $\mathbb{A}$ as

$$Z(x, t; \alpha, \beta) := \int_{\mathbb{A}} \chi(-\xi \cdot x) \, e^{-t\left(|\xi_\infty|_\infty^\beta + \|\xi_f\|^\alpha\right)} d\xi_{\mathbb{A}}, \quad x \in \mathbb{A}, \ t > 0.$$

Since $e^{-t|\xi_\infty|_\infty^\beta} \in L^1(\mathbb{R}, d\xi_\infty)$, $e^{-t\|\xi_f\|^\alpha} \in L^1\left(\mathbb{A}_f, d\xi_{\mathbb{A}_f}\right)$, cf. [38, Property 2.2] and Proposition 100 , and $d\xi_{\mathbb{A}} = d\xi_\infty d\xi_{\mathbb{A}_f}$, we have

$$Z(x, t; \alpha, \beta) = \mathcal{F}^{-1}\left(e^{-t|\xi_\infty|_\infty^\beta}\right) \mathcal{F}^{-1}\left(e^{-t\|\xi_f\|^\alpha}\right) = Z(x_\infty, t; \beta) Z\left(x_f, t; \alpha\right).$$

$$(4.53)$$

For $t > 0$ fixed, we use the notation $Z_t(x; \alpha, \beta)$ instead of $Z(x, t; \alpha, \beta)$.

Theorem 120 *The adelic heat kernel on* $\mathbb{A}$ *possesses the following properties*

   (i) $Z(x, t; \alpha, \beta) \geq 0$ *for any* $t > 0$;
  (ii) $\int_{\mathbb{A}} Z(x, t; \alpha, \beta) \, dx_{\mathbb{A}} = 1$ *for any* $t > 0$;
 (iii) $Z_t(x; \alpha, \beta) \in L^1(\mathbb{A})$ *for any* $t > 0$;
 (iv) $Z_t(x; \alpha, \beta) * Z_{t'}(x; \alpha, \beta) = Z_{t+t'}(x; \alpha, \beta)$ *for any* $t, t' > 0$;
  (v) $\lim_{t \to 0+} Z(x, t; \alpha, \beta) = \delta(x_\infty) \times \delta\left(x_f\right) = \delta(x)$ *in* $\mathcal{D}'(\mathbb{A})$;
 (vi) $Z_t(x; \alpha, \beta)$ *is a uniformly continuous function for any fixed* $t > 0$;
(vii) $Z(x, t; \alpha, \beta)$ *is uniformly continuous in* t, *i.e.* $Z(x, t; \alpha, \beta)$ *belongs to* $C((0, \infty), C(\mathbb{A}))$ *or* $\lim_{t' \to t} \max_{x \in \mathbb{A}} |Z(x, t; \alpha, \beta) - Z(x, t'; \alpha, \beta)| = 0$ *for any* $t > 0$.

Proof The statement follows from (4.53) and the corresponding properties for $Z(x_\infty, t; \beta)$ and $Z\left(x_f, t; \alpha\right)$, see [38, Section 2] and Theorem 105. ∎

4.9 Markov Processes on $\mathbb{A}$

Let $\mathcal{B}(\mathbb{A})$ denote the σ-algebra of the Borel sets of $(\mathbb{A}, \rho_{\mathbb{A}})$. Along this section we suppose that $\alpha > 1$ and $\beta \in (0, 2]$ are fixed parameters. We set

$$p(t, x, y; \alpha, \beta) := Z(x - y, t; \alpha, \beta) \qquad \text{for } t > 0, x, y \in \mathbb{A}.$$

Note that

$$p(t, x, y; \alpha, \beta) = Z(x_\infty - y_\infty, t; \beta) Z\left(x_f - y_f, t; \alpha\right)$$

$$=: p(t, x_\infty, y_\infty; \beta) \, p\left(t, x_f, y_f; \alpha\right),$$

where $p\,(t, x_\infty, y_\infty; \beta) \;=\; Z(x_\infty - y_\infty, t; \beta)$ and $p\,\big(t, x_f, y_f; \alpha\big) := p\,\big(t, x_f, y_f\big) = Z(x_f - y_f, t; \alpha)$. We also define for $x_\infty, y_\infty \in \mathbb{R}$ and $B_\infty \in \mathcal{B}\,(\mathbb{R})$

$$P\,(t, x_\infty, B_\infty; \beta) := \begin{cases} \int_{B_\infty} p\,(t, x_\infty, y_\infty; \beta)\,dy_\infty & \text{for } t > 0 \\ 1_{B_\infty}\,(x_\infty) & \text{for } t = 0 \end{cases}$$

and for $x, y \in \mathbb{A}$ and $B \in \mathcal{B}\,(\mathbb{A})$

$$P\,(t, x, B; \alpha, \beta) := \begin{cases} \int_B p\,(t, x_\infty, y_\infty; \beta)\,p\,\big(t, x_f, y_f; \alpha\big)\,dy_\infty dy_{\mathbb{A}_f} & \text{for } t > 0 \\ 1_B\,(x) & \text{for } t = 0. \end{cases}$$

Lemma 121 *With the above notation the following assertions hold:*

(i) $p\,(t, x, y; \alpha, \beta)$ *is a normal transition density;*
(ii) $P\,(t, x, B; \alpha, \beta)$ *is a normal transition function.*

Proof The statement follows from the corresponding properties for the functions $p\,(t, x_\infty, y_\infty; \beta)$ and $p\,\big(t, x_f, y_f\big)$, see Lemma 106. ∎

Lemma 122 *The transition function $P\,(t, x, B; \alpha, \beta)$ satisfies the following two conditions:*

(i) for each $u \geq 0$ and a compact B

$$\lim_{x \to \infty} \sup_{t \leq u} P\,(t, x, B; \alpha, \beta) = 0; \qquad [\text{Condition } L(B)]$$

(ii) for each $\epsilon > 0$ and a compact B

$$\lim_{t \to 0+} \sup_{x \in B} P\Big(t, x, \mathbb{A} \setminus \overset{\circ}{B}_\epsilon\,(x)\,; \alpha, \beta\Big) = 0, \qquad [\text{Condition } M(B)]$$

where $\overset{\circ}{B}_\epsilon\,(x) := \{y \in \mathbb{A}; \rho_{\mathbb{A}}\,(x, y) < \epsilon\}$.

Proof

(i) Note that there exist compact subsets $K_\infty \subset \mathbb{R}$ and $K_f \subset \mathbb{A}_f$ such that $B \subset K_\infty \times K_f$. Then

$$P(t, x, B; \alpha, \beta) \leq P(t, x_\infty, K_\infty; \beta) P(t, x_f, K_f).$$

Since $\rho_{\mathbb{A}}\,(0, x) \to \infty$ we have either $\rho\,(0, x_f) \to \infty$ or $|x_\infty|_\infty \to \infty$. Therefore it is sufficient to show that

$$\lim_{x_f \to \infty} \sup_{t \leq u} P(t, x_f, K_f) = 0 \qquad (4.54)$$

and

$$\lim_{x_\infty \to \infty} \sup_{t \le u} P(t, x_\infty, K_\infty; \beta) = 0. \tag{4.55}$$

The equality (4.54) follows from Lemma 107. By [38, (2.2)]

$$Z(t, x_\infty; \beta) \le \frac{Ct^{\frac{1}{\beta}}}{t^{\frac{2}{\beta}} + x_\infty^2} \qquad \text{for } t > 0, x_\infty \in \mathbb{R}. \tag{4.56}$$

Then

$$P(t, x_\infty, K_\infty; \beta) = \int_{K_\infty} Z(t, x_\infty - y_\infty; \beta) dy_\infty$$

$$\le Ct^{\frac{1}{\beta}} \int_{K_\infty} \frac{1}{t^{\frac{2}{\beta}} + (x_\infty - y_\infty)^2} dy_\infty.$$

As $x_\infty \to \infty$ we have dist $(x_\infty, K_\infty) \to \infty$ and $|x_\infty - y_\infty| \ge \text{dist}(x_\infty; K_\infty)$ for any $y_\infty \in K_\infty$ and

$$\frac{1}{t^{\frac{2}{\beta}} + (x_\infty - y_\infty)^2} \le \frac{1}{\text{dist}^2(x_\infty; K_\infty)}.$$

Hence

$$\lim_{x_\infty \to \infty} \sup_{t \le u} P(t, x_\infty, K_\infty; \beta) \le \lim_{x_\infty \to \infty} Cu^{\frac{1}{\beta}} \int_{K_\infty} \frac{1}{\text{dist}^2(x_\infty; K_\infty)} dy_\infty = 0.$$

(ii) Since $\overset{\circ}{B}_\epsilon (x) \supseteq \overset{\circ}{B}_{\frac{\epsilon}{2}} (x_\infty) \times B_{\frac{\epsilon}{2}} (x_f)$, where

$$\overset{\circ}{B}_{\frac{\epsilon}{2}} (x_\infty) = \left\{ y_\infty \in \mathbb{R}; |x_\infty - y_\infty| < \frac{\epsilon}{2} \right\}$$

and $B_{\frac{\epsilon}{2}} (x_f)$ is given by (4.8), we have $\mathbb{A} \setminus \overset{\circ}{B}_\epsilon (x) \subseteq (\mathbb{R} \times \mathbb{A}_f) \setminus \left(\overset{\circ}{B}_{\frac{\epsilon}{2}} (x_\infty) \times B_{\frac{\epsilon}{2}} (x_f) \right) \subseteq \left((\mathbb{R} \setminus \overset{\circ}{B}_{\frac{\epsilon}{2}} (x_\infty)) \times \mathbb{A}_f \right) \cup \left(\mathbb{R} \times (\mathbb{A}_f \setminus B_{\frac{\epsilon}{2}} (x_f)) \right)$ and with the use of [38, (2.1)] and Theorem 105 (ii) we obtain

$$P\left(t, x, \mathbb{A} \setminus \overset{\circ}{B}_\epsilon (x); \alpha, \beta\right) \le \left(\int_{|x_\infty - y_\infty|_\infty \ge \frac{\epsilon}{2}} p(t, x_\infty, y_\infty; \beta) dy_\infty \right)$$

$$+ \left(\int_{\rho(x_f, y_f) > \frac{\epsilon}{2}} p(t, x_f, y_f) dy_{\mathbb{A}_f} \right)$$

$$\le P\left(t, x_\infty, \mathbb{R} \setminus \overset{\circ}{B}_{\frac{\epsilon}{2}} (x); \beta\right) + P\left(t, x_f, \mathbb{A}_f \setminus B_{\frac{\epsilon}{2}} (x_f)\right).$$

Now the result follows from Lemma 107 and the inequality

$$P\left(t, x_\infty, \mathbb{R} \setminus \overset{\circ}{B}_{\frac{\varepsilon}{2}}(x)\right) = \int_{|y_\infty| \geq \frac{\varepsilon}{2}} Z(t, y_\infty; \beta) \, dy_\infty$$

$$\leq C \int_{|y_\infty| \geq \frac{\varepsilon}{2}} \frac{t^{\frac{1}{\beta}}}{t^{\frac{2}{\beta}} + y_\infty^2} \, dy_\infty$$

$$= C \int_{|z_\infty| \geq \frac{\varepsilon}{2} t^{-1/\beta}} \frac{1}{1 + z_\infty^2} \, dz_\infty \to 0 \qquad \text{as } t \to +0. \qquad \blacksquare$$

Theorem 123 $Z(x, t; \alpha, \beta)$ *with* $\alpha > 1$ *and* $\beta \in (0, 2]$ *is the transition density of a time- and space homogenous Markov process which is bounded, right-continuous and has no discontinuities other than jumps.*

Proof The result follows from [39, Theorem 3.6], Remark 96 (ii) and Lemmas 121, 122. $\qquad \blacksquare$

4.10 Cauchy Problem for Parabolic Type Equations on $\mathbb{A}$

In this section we study Cauchy problems for parabolic type equations on $\mathbb{A}$ and present analogues of the results of Sects. 4.7.1, 4.7.2 and 4.7.4.

4.10.1 Homogeneous Equations

Consider the following Cauchy problem

$$\begin{cases} \dfrac{\partial u(x, t)}{\partial t} + \boldsymbol{D}^{\alpha, \beta} u(x, t) = 0, & x \in \mathbb{A}, \ t \in [0, +\infty) \\[2mm] u(x, 0) = u_0(x), & u_0(x) \in Dom\left(\boldsymbol{D}^{\alpha, \beta}\right), \end{cases} \qquad (4.57)$$

where $\alpha > 1$, $\beta \in (0, 2]$, $\boldsymbol{D}^{\alpha, \beta}$ is the pseudodifferential operator defined by (4.38) with the domain given by (4.41) and $u : \mathbb{A} \times [0, \infty) \to \mathbb{C}$ is an unknown function. We say that a function $u(x, t)$ is a *solution of* (4.57), if $u \in C\left([0, \infty), Dom(\boldsymbol{D}^{\alpha, \beta})\right) \cap C^1\left([0, \infty), L^2(\mathbb{A})\right)$ and if u satisfies equation (4.57) for all $t \geq 0$.

As in Sect. 4.7 we understand the notions of continuity, differentiability and equalities in the sense of $L^2(\mathbb{A})$.

We first consider Cauchy problem (4.57) with the initial value from $\mathcal{D}(\mathbb{A})$. We define the function

$$u(x,t) := u(x,t;\alpha,\beta) = Z_t(x;\alpha,\beta) * u_0(x) = Z_t(x) * u_0(x), \qquad t \geq 0, \qquad (4.58)$$

where $Z_0 * u_0 = (Z_t * u_0)\big|_{t=0} := u_0$. Note that such definition is consistent with Theorem 120 (v). Since $Z_t(x) \in L^1(\mathbb{A})$ for $t > 0$ and $u_0(x) \in \mathcal{D}(\mathbb{A}) \subset L^\infty(\mathbb{A})$, the convolution exists and is a continuous function, see Theorem 120 (ii), [100, Theorem 1.1.6].

Lemma 124 *Let $u_0 \in \mathcal{D}(\mathbb{A})$ and u is defined by (4.58). Then u belongs to $C\big([0,\infty), Dom(\mathbf{D}^{\alpha,\beta})\big)$ and*

$$\mathbf{D}^{\alpha,\beta} u = \mathcal{F}_{\xi \to x}^{-1} \left(\left(|\xi_\infty|_\infty^\beta + \|\xi_f\|^\alpha \right) e^{-t\left(|\xi_\infty|_\infty^\beta + \|\xi_f\|^\alpha \right)} \mathcal{F}_{x \to \xi} u_0 \right) \qquad (4.59)$$

for $t \geq 0$.

Proof We first verify that $u(\cdot,t) \in Dom(\mathbf{D}^{\alpha,\beta})$ for $t \geq 0$. Without loss of generality we may assume that $u_0(x) = u_\infty(x_\infty) u_f(x_f)$ with $u_\infty \in \mathcal{D}(\mathbb{R})$ and $u_f \in \mathcal{D}(\mathbb{A}_f)$. Since

$$\mathcal{F}_{x \to \xi} u(x,t) = e^{-t\left(|\xi_\infty|_\infty^\beta + \|\xi_f\|^\alpha \right)} \hat{u}_\infty(\xi_\infty) \hat{u}_f(\xi_f),$$

we have

$$\left\| \left(|\xi_\infty|_\infty^\beta + \|\xi_f\|^\alpha \right) \mathcal{F}_{x \to \xi} u \right\|_{L^2(\mathbb{A})}$$

$$\leq \left\| \|\xi_f\|^\alpha e^{-t\|\xi_f\|^\alpha} \hat{u}_f(\xi_f) \right\|_{L^2(\mathbb{A}_f)} \cdot \left\| e^{-t|\xi_\infty|_\infty^\beta} \hat{u}_\infty(\xi_\infty) \right\|_{L^2(\mathbb{R})}$$

$$\quad + \left\| e^{-t\|\xi_f\|^\alpha} \hat{u}_f(\xi_f) \right\|_{L^2(\mathbb{A}_f)} \cdot \left\| |\xi_\infty|_\infty^\beta e^{-t|\xi_\infty|_\infty^\beta} \hat{u}_\infty(\xi_\infty) \right\|_{L^2(\mathbb{R})}$$

$$\leq \left\| \|\xi_f\|^\alpha \hat{u}_f(\xi_f) \right\|_{L^2(\mathbb{A}_f)} \cdot \left\| \hat{u}_\infty \right\|_{L^2(\mathbb{R})} + \left\| \hat{u}_f \right\|_{L^2(\mathbb{A}_f)} \cdot \left\| |\xi_\infty|_\infty^\beta \hat{u}_\infty(\xi_\infty) \right\|_{L^2(\mathbb{R})}$$

$$= \left\| \mathbf{D}^\alpha u_f \right\|_{L^2(\mathbb{A}_f)} \left\| u_\infty \right\|_{L^2(\mathbb{R})} + \left\| u_f \right\|_{L^2(\mathbb{A}_f)} \left\| \mathbf{D}^\beta u_\infty \right\|_{L^2(\mathbb{R})},$$

where we used the equality $d\xi_\mathbb{A} = d\xi_{\mathbb{A}_f} d\xi_\infty$ and the Parseval-Steklov formula. Therefore $u(x,t) \in Dom(\mathbf{D}^{\alpha,\beta})$ for $t \geq 0$ and formula (4.59) holds.

To verify the continuity, assume again that $u_0(x) = u_\infty(x_\infty) u_f(x_f)$ with $u_\infty \in \mathcal{D}(\mathbb{R})$, $u_f \in \mathcal{D}(\mathbb{A}_f)$. With the use of the Parseval-Steklov equality and the mean value theorem we obtain

$$\lim_{t' \to t} \left\| u(x,t) - u(x,t') \right\|_{L^2(\mathbb{A})}$$

$$= \lim_{t' \to t} \left\| \left(e^{-t\left(|\xi_\infty|_\infty^\beta + \|\xi_f\|^\alpha \right)} - e^{-t'\left(|\xi_\infty|_\infty^\beta + \|\xi_f\|^\alpha \right)} \right) \hat{u}_\infty(\xi_\infty) \hat{u}_f(\xi_f) \right\|_{L^2(\mathbb{A})}$$

$$= \lim_{t' \to t} \left\| (t - t') \left(|\xi_\infty|_\infty^\beta + \|\xi_f\|^\alpha \right) e^{-\tilde{t} \cdot \left(|\xi_\infty|_\infty^\beta + \|\xi_f\|^\alpha \right)} \hat{u}_\infty (\xi_\infty) \, \hat{u}_f \left(\xi_f \right) \right\|_{L^2(\mathbb{A})}$$

$$\leq \lim_{t' \to t} |t - t'| \cdot \left\| \left(|\xi_\infty|_\infty^\beta + \|\xi_f\|^\alpha \right) \hat{u}_\infty (\xi_\infty) \, \hat{u}_f \left(\xi_f \right) \right\|_{L^2(\mathbb{A})}$$

$$\leq \left(\left\| D^\alpha u_f \right\|_{L^2(\mathbb{A}_f)} \left\| u_\infty \right\|_{L^2(\mathbb{R})} + \left\| u_f \right\|_{L^2(\mathbb{A}_f)} \left\| D^\beta u_\infty \right\|_{L^2(\mathbb{R})} \right) \lim_{t' \to t} |t - t'| = 0,$$

where $\tilde{t} = \tilde{t}\left(|\xi_\infty|_\infty^\beta + \|\xi_f\|^\alpha \right)$ is a point between t and t'.   ∎

Lemma 125 *Let $u_0 \in \mathcal{D}(\mathbb{A})$ and $u(x, t)$, $t \geq 0$ is defined by (4.58). Then $u(x, t)$ is continuously differentiable in time for $t \geq 0$ and the derivative is given by*

$$\frac{\partial u}{\partial t}(x, t) = -\mathcal{F}_{\xi \to x}^{-1} \left(\left(|\xi_\infty|_\infty^\beta + \|\xi_f\|^\alpha \right) e^{-t \left(|\xi_\infty|_\infty^\beta + \|\xi_f\|^\alpha \right)} \mathcal{F}_{x \to \xi} u_0 \right). \tag{4.60}$$

Proof Assume that $u_0(x) = u_\infty(x_\infty) u_f(x_f)$ with $u_\infty \in \mathcal{D}(\mathbb{R})$, $u_f \in \mathcal{D}(\mathbb{A}_f)$. By reasoning as in the proofs of Lemmas 110 and 124, we have

$$\lim_{t \to t_0} \left\| \frac{\hat{u}(\xi, t) - \hat{u}(\xi, t_0)}{t - t_0} + \left(|\xi_\infty|_\infty^\beta + \|\xi_f\|^\alpha \right) e^{-t \left(|\xi_\infty|_\infty^\beta + \|\xi_f\|^\alpha \right)} \mathcal{F}_{x \to \xi} u_0 \right\|_{L^2(\mathbb{A})}$$

$$= \lim_{t \to t_0} |t - t_0| \cdot \left\| \left(|\xi_\infty|_\infty^\beta + \|\xi_f\|^\alpha \right)^2 e^{-\tilde{t} \cdot \left(|\xi_\infty|_\infty^\beta + \|\xi_f\|^\alpha \right)} \mathcal{F}_{x \to \xi} u_0 \right\|_{L^2(\mathbb{A})}$$

$$\leq \lim_{t \to t_0} |t - t_0| \cdot \left\| \left(|\xi_\infty|_\infty^\beta + \|\xi_f\|^\alpha \right)^2 \mathcal{F}_{x \to \xi} u_0 \right\|_{L^2(\mathbb{A})}$$

$$\leq \left(\left\| D^{2\alpha} u_f \right\| \cdot \left\| u_\infty \right\| + 2 \left\| D^\alpha u_f \right\| \cdot \left\| D^\beta u_\infty \right\| + \left\| u_f \right\| \cdot \left\| D^{2\beta} u_\infty \right\| \right) \lim_{t' \to t} |t - t'| = 0,$$

where we have used the fact that $\mathcal{D}(\mathbb{R}) \subset Dom(D^\beta)$ for any $\beta > 0$ and $\mathcal{D}(\mathbb{A}_f) \subset Dom(D^\alpha)$ for any $\alpha > 0$.

To verify the continuity of $\frac{\partial u}{\partial t}(x, t)$, we proceed similarly:

$$\lim_{t \to t_0} \left\| \frac{\partial u}{\partial t}(x, t) - \frac{\partial u}{\partial t}(x, t_0) \right\|_{L^2(\mathbb{A})}$$

$$= \lim_{t \to t_0} |t_0 - t| \left\| \left(|\xi_\infty|_\infty^\beta + \|\xi_f\|^\alpha \right)^2 e^{-\tilde{t} \left(|\xi_\infty|_\infty^\beta + \|\xi_f\|^\alpha \right)} \hat{u}_\infty (\xi_\infty) \, \hat{u}_f \left(\xi_f \right) \right\|_{L^2(\mathbb{A})}$$

$$\leq \lim_{t \to t_0} |t_0 - t| \left\| \left(|\xi_\infty|_\infty^\beta + \|\xi_f\|^\alpha \right)^2 \hat{u}_\infty (\xi_\infty) \, \hat{u}_f \left(\xi_f \right) \right\|_{L^2(\mathbb{A})} = 0.$$

where we used the mean value theorem with a point $\tilde{t}$ between t and t_0.   ∎

As an immediate consequence from Lemmas 124 and 125 we obtain

Proposition 126 *Let the function $u_0 \in \mathcal{D}(\mathbb{A})$. Then the function $u(x, t)$ defined by (4.58) is a solution of Cauchy problem (4.57).*

Consider the operator $T(t; \alpha, \beta)$, $t \geq 0$ of convolution with the adelic heat kernel

$$T(t; \alpha, \beta)u = Z_t * u. \tag{4.61}$$

As in Sect. 4.7, the convolution $Z_t * u$ is a continuous function of x for $t > 0$ and any $u \in L^2(\mathbb{A})$ and the operator $T(t; \alpha, \beta) : L^2(\mathbb{A}) \to L^2(\mathbb{A})$ is bounded.

By reasoning as in the proof of Theorem 114, we obtain

Theorem 127 *Let $\alpha > 1$ and $\beta \in (0, 2]$. Then the following assertions hold.*

(i) *The operator $-\boldsymbol{D}^{\alpha,\beta}$ generates a C_0 semigroup $\left(\mathcal{T}(t; \alpha, \beta)\right)_{t\geq 0}$. The operator $\mathcal{T}(t; \alpha, \beta)$ coincides for each $t \geq 0$ with the operator $T(t; \alpha, \beta)$ given by (4.61).*
(ii) *Cauchy problem (4.57) is well-posed and its solution is given by $u(x, t) = Z_t * u_0$, $t \geq 0$.*

4.10.2 Non Homogeneous Equations

Consider the following Cauchy problem

$$\begin{cases} \dfrac{\partial u(x, t)}{\partial t} + \boldsymbol{D}^{\alpha,\beta} u(x, t) = f(x, t), & x \in \mathbb{A}, \ t \in [0, T], \ T > 0 \\[2mm] u(x, 0) = u_0(x), & u_0(x) \in Dom\left(\boldsymbol{D}^{\alpha,\beta}\right). \end{cases} \tag{4.62}$$

We say that a function $u(x, t)$ is a *solution* of (4.62), if u belongs to $C\left([0, T], Dom(\boldsymbol{D}^{\alpha,\beta})\right) \cap C^1\left([0, T], L^2(\mathbb{A})\right)$ and if u satisfies equation (4.62) for $t \in [0, T]$.

Theorem 128 *Let $\alpha > 1$, $\beta \in (0, 2]$ and let $f \in C\left([0, T], L^2(\mathbb{A})\right)$. Assume that at least one of the following conditions is satisfied:*

(i) $f \in L^1\left((0, T), Dom(\boldsymbol{D}^{\alpha,\beta})\right)$;
(ii) $f \in W^{1,1}\left((0, T), L^2(\mathbb{A})\right)$.

Then Cauchy problem (4.62) has a unique solution given by

$$u(x, t) = \int_{\mathbb{A}} Z(x - y, t; \alpha, \beta) u_0(y) \, dy_{\mathbb{A}}$$

$$+ \int_0^t \left\{ \int_{\mathbb{A}} Z(x - y, t - \tau; \alpha, \beta) f(y, \tau) \, dy_{\mathbb{A}} \right\} d\tau.$$

Proof With the use of Theorem 127 the proof follows from well-known results of the semigroup theory, see e.g. [7, Proposition 3.1.16], [24, Proposition 4.1.6]. ∎

Chapter 5
Fundamental Solutions for Pseudodifferential Operators, and Equations of Schrödinger Type

5.1 Introduction

This chapter aims to explore the connections between local zeta functions and fundamental solutions for pseudo-differential operators over p-adic fields. In the 50s Gel'fand and Shilov showed that fundamental solutions for certain types of partial differential operators with constant coefficients can be obtained by using local zeta functions [49]. The existence of fundamental solutions for general differential operators with constant coefficients was established by Atiyah [15] and Bernstein [18] using local zeta functions. A similar program can be carried out in the p-adic setting. In this chapter, we give a detailed proof of the existence of a fundamental solution for a pseudo-differential operator with a symbol of the form $|f|_p^\alpha$, with f an arbitrary polynomial, see Theorem 134. This result was proved in [123], see also [120], by Zúñiga-Galindo, here we present a complete proof, including a review of the method of analytic continuation of Gel'fand-Shilov, see [49].

This chapter is organized as follows. In Sects. 5.2–5.3, we review the method of analytic continuation of Gel'fand-Shilov, and the basic aspects of the local zeta functions. In Sect. 5.4, we prove the existence of a fundamental solution E_β for a pseudodifferential operator $\mathbf{f}(\partial, \beta)$, see Theorem 134. For small β, there exists a simple formula for E_β, see Theorem 137. In Sect. 5.5, we compute fundamental solutions for quasielliptic operators, see Theorem 142 and Corollary 143. In Sect. 5.6, we compute fundamental solutions for Schrödinger-type pseudodifferential operators, see Theorem 146. We study certain homogeneous equations attached to these operators, see Theorem 149, and finally we study some initial value problems for Schrödinger-type pseudodifferential equations, see Theorem 150.

© Springer International Publishing AG 2016
W.A. Zúñiga-Galindo, *Pseudodifferential Equations Over
Non-Archimedean Spaces*, Lecture Notes in Mathematics 2174,
DOI 10.1007/978-3-319-46738-2_5

5.2   The Method of Analytic Continuation of Gel'fand-Shilov

In this section, we review the method of analytic continuation of Gel'fand-Shilov, see [49, Appendix B] or [57, pp. 65–67].

Let U be a non-empty, open subset of $\mathbb{C}$. Assume that

$$U \to \mathcal{D}'$$
$$s \to T_s$$

is an $\mathcal{D}'$-valued function on U. We say that T_s *is continuous*, respectively *holomorphic*, on U if (T_s, ϕ) is continuous, respectively holomorphic, on U for every ϕ in $\mathcal{D}$. Let X be a continuous curve of finite length in U (i.e. continuous on $X \cap U$), then

$$(R, \phi) := \int_X (T_s, \phi)\, ds$$

is defined for every ϕ in $\mathcal{D}$ and defines an element of $\mathcal{D}'$. We denote this distribution as

$$R = \int_X T_s ds.$$

Suppose that T_s is holomorphic on U, and that the puncture disc

$$\{s \in \mathbb{C}; 0 < |s - s_0| \leq r\}$$

is contained in U for some $s_0 \in \mathbb{C}$ (not necessarily in U) and some $r > 0$. Then (T_s, ϕ) can be expanded in its Laurent series:

$$(T_s, \phi) = \sum_{k \in \mathbb{Z}} (a_k, \phi)\, (s - s_0)^k,$$

where

$$a_k = \frac{1}{2\pi\sqrt{-1}} \int_{|s - s_0| \leq r} \frac{T_s}{(s - s_0)^{k+1}} ds \tag{5.1}$$

is in $\mathcal{D}'$. Now, since $\mathcal{D}'$ is complete

$$T_s = \sum_{k \in \mathbb{Z}} a_k\, (s - s_0)^k$$

gives an element of $\mathcal{D}'$, the *Laurent expansion of* T_s at s_0. We say that s_0 is a pole of T_s if $a_k \neq 0$ (in $\mathcal{D}'$) only for a finite number of $k < 0$.

5.3 Igusa's Local Zeta Functions

We set for $a > 0$ and $s \in \mathbb{C}$, $a^s := e^{s \ln a}$. Let $f(\xi) \in \mathbb{Q}_p[\xi_1, \dots, \xi_n]$ be a non-constant polynomial. *The p-adic complex power* $|f|_p^s$ associated to f (also called *the Igusa zeta function of f*) is the distribution

$$\left(|f|_p^s, \phi\right) := \int_{\mathbb{Q}_p^n \smallsetminus f^{-1}(0)} \phi(\xi) |f(\xi)|_p^s \, d^n\xi, \ s \in \mathbb{C}, \ \mathrm{Re}(s) > 0.$$

The Igusa local zeta functions are connected with the number of solutions of polynomial congruences $\bmod p^m$ and with exponential sums $\bmod p^m$. There are many intriguing conjectures connecting the poles of local zeta functions with topology of complex singularities, see e.g. [30, 57]. The local zeta functions can be defined on any locally compact field K, i.e. for $\mathbb{R}$, $\mathbb{C}$, or finite extensions of $\mathbb{Q}_p$ or of $\mathbb{F}_p((t))$, with $\mathbb{F}_p$ the finite field with p elements. In the Archimedean case $K = \mathbb{R}$ or $\mathbb{C}$, the study of local zeta functions was initiated by I.M. Gel'fand and G.E. Shilov [49]. The main motivation was the construction of fundamental solutions for partial differential operators with constant coefficients. Indeed, the meromorphic continuation of the local zeta functions imply the existence of fundamental solutions, this fact was established, independently, by Atiyah [15] and Bernstein [18], see also [57, Theorem 5.5.1 and Corollary 5.5.1]. On the other hand, in the middle 60s, A. Weil initiated the study of local zeta functions, in the Archimedean and non Archimedean settings, in connection with the Poisson-Siegel formula. In the 70s, Igusa developed a uniform theory for local zeta functions over local fields of characteristic zero [57].

Theorem 129 (Igusa [57, Theorem 8.2.1]) *For* $\mathrm{Re}(s) > 0$, $|f|_p^s$ *defines a* $\mathcal{D}'$-*valued holomorphic function, and it has a meromorphic continuation to the whole complex plane such that* $\left(|f|_p^s, \phi\right)$ *is a rational function of* p^{-s}. *More precisely, there exists a finite collection of pairs of no-negative integers* $\{(N_E, v_E) \in \mathbb{N} \times \mathbb{N} \smallsetminus \{0\} ; E \in \mathcal{T}\}$ *depending only on* f, *such that*

$$\prod_{E \in \mathcal{T}} \left(1 - p^{-v_E - N_E s}\right) \cdot |f|_p^s$$

becomes a $\mathcal{D}'$-*valued holomorphic function on the whole complex plane.*

Remark 130 (i) Notice that the poles of $|f|_p^s$ have the form $s = -\frac{v_E}{N_E} + \frac{2\pi i}{N_E \ln p} \mathbb{Z}$.
(ii) Let $f : \mathbb{Q}_p^n \longrightarrow \mathbb{Q}_p$ be a polynomial mapping satisfying $f(0) = 0$. Let $\{(N_E, v_E) \in \mathbb{N} \times \mathbb{N} \smallsetminus \{0\} ; E \in \mathcal{T}\}$ be as in Theorem 129. Set $\lambda := \lambda(f) = \min_E \frac{v_E}{N_E}$. Then $-\lambda$ is the real part of a pole of $\left(|f|_p^s, \phi\right)$ for some $\phi \in \mathcal{D}$, cf. [110, Theorem 2.7] or [58]. Notice that $-\lambda < 0$. This result implies that

$$\int_{\mathbb{Q}_p^n \smallsetminus f^{-1}(0)} \frac{\phi(\xi)}{|f(\xi)|_p^\beta} d^n\xi < \infty \text{ for any } \beta \text{ satisfying } 0 < \beta < \lambda(f).$$

From these observations and by applying Theorem 129, we get the following result:

Corollary 131 *With the above notation,*

$$\left(|f|_p^s, \phi\right) = \int\limits_{\mathbb{Q}_p^n \setminus f^{-1}(0)} \phi\left(\xi\right) |f\left(\xi\right)|_p^s d^n\xi, \text{ for } \mathrm{Re}(s) > -\lambda\left(f\right),$$

defines a $\mathcal{D}'$-value holomorphic function.

Remark 132 We call a $\mathcal{D}'$-valued holomorphic function a holomorphic distribution on $\mathcal{D}$.

5.4 Fundamental Solutions for Pseudodifferential Operators

Let $f\left(\xi\right) \in \mathbb{Q}_p[\xi_1, \ldots, \xi_n]$ be an arbitrary non-constant polynomial. *A pseudodifferential operator with symbol* $|f\left(\xi\right)|_p^\beta$, $\beta > 0$, is an extension of an operator of the form

$$\left(f\left(\partial, \beta\right)\phi\right)(x) = \mathcal{F}_{\xi \to x}^{-1}\left(|f\left(\xi\right)|_p^\beta \mathcal{F}_{x \to \xi}\phi\right), \text{ for } \phi \in \mathcal{D}.$$

Definition 133 Let $E_\beta \in \mathcal{D}'$, with $\beta > 0$. We say that E_β is *a fundamental solution* for

$$f\left(\partial, \beta\right) u = \phi, \text{ with } \phi \in \mathcal{D}, \qquad (5.2)$$

if $u = E_\beta * \phi$ is a solution of (5.2) in $\mathcal{D}'$.

At this point, it is important to mention that we cannot use the standard definition of fundamental solution, i.e. $f\left(\partial, \beta\right) E_\beta = \delta$, because $\mathcal{D}$ is not invariant under the action of $f\left(\partial, \beta\right)$.

Theorem 134 *There exists a fundamental solution for $f\left(\partial, \beta\right) u = \phi$, with $\phi \in \mathcal{D}$.*

Proof We first note that the existence of a fundamental solution for $f\left(\partial, \beta\right) u = \phi$, with $\phi \in \mathcal{D}$, is equivalent to the existence of a distribution $T_\beta = \mathcal{F}E_\beta$ satisfying

$$|f\left(\xi\right)|_p^\beta T_\beta = 1 \text{ in } \mathcal{D}', \qquad (5.3)$$

where $(1, \phi) = \int \phi d^n x$. Indeed, assume that (5.3) holds. Then for any $\mathcal{F}\phi \in \mathcal{D}$, $|f\left(\xi\right)|_p^\beta T_\beta \mathcal{F}\phi = \mathcal{F}\phi$ in $\mathcal{D}'$. The product $|f\left(\xi\right)|_p^\beta T_\beta \mathcal{F}\phi$ is well-defined as a commutative and associative product, cf. [105, Theorem 3.19], thus,

$$|f\left(\xi\right)|_p^\beta T_\beta \mathcal{F}\phi = |f\left(\xi\right)|_p^\beta \left(T_\beta \mathcal{F}\phi\right) = |f\left(\xi\right)|_p^\beta \mathcal{F}\left(E_\beta * \phi\right) = |f\left(\xi\right)|_p^\beta \mathcal{F}u = \mathcal{F}\phi$$

in $\mathcal{D}'$, i.e. $f(\partial, \beta)u = \phi$ in $\mathcal{D}'$. The other implication is established in a similar form.

Claim A $|f|_p^{s+\beta} = |f|_p^{\beta}\,|f|_p^{s}$ in $\mathcal{D}'$, for any $s \in C$.

We show the existence of a solution for the division problem (5.3). By Theorem 129, $|f|_p^{s}$ has a meromorphic continuation to the complex plane, then let

$$|f|_p^{s} = \sum_{k\in\mathbb{Z}} a_k\,(s+\beta)^k$$

be the Laurent expansion of $|f|_p^{s}$ at $-\beta$, with $a_k \in \mathcal{D}'$. Since the real parts of the poles of $|f|_p^{s}$ are negative rational numbers, by Claim A, $|f|_p^{s+\beta} = |f|_p^{s}\,|f|_p^{\beta}$ is a holomorphic distribution at $s = -\beta$. Therefore $|f|_p^{\beta}\,a_k = 0$ for all $k < 0$ (notice that by (5.1) and Claim A, $|f|_p^{\beta}\,a_k \in \mathcal{D}'$), and

$$|f|_p^{s+\beta} = |f|_p^{\beta}\,a_0 + \sum_{k=1}^{\infty} |f|_p^{\beta}\,a_k\,(s+\beta)^k. \tag{5.4}$$

By applying the dominated convergence theorem in (5.4):

$$\lim_{s\to-\beta}\left(|f|_p^{s+\beta},\phi\right) = (1,\phi) = |f|_p^{\beta}\,a_0,$$

hence $T_\beta = a_0$, and $E_\beta = \mathcal{F}^{-1}a_0$.

Proof of Claim A We recall that the product $|f|_p^{s}\,|f|_p^{\beta}$ is defined by

$$\left(|f|_p^{s}\,|f|_p^{\beta},\varphi\right) = \lim_{j\to+\infty}\left(|f|_p^{s},\left\{|f|_p^{\beta}*\delta_j\right\}\varphi\right) \quad \text{for } \mathrm{Re}(s) > 0,$$

where $\delta_j(x) = p^{nj}\Omega\left(p^j\,\|x\|_p\right)$, if the limit exists for all $\varphi \in \mathcal{D}$. By using the fact that if $x \in \mathbb{Q}_p^n \setminus f^{-1}(0)$, then $\left|f\left(x+p^j\tilde{y}\right)\right|_p = |f(x)|_p$ for j big enough (depending on x) and for any $\tilde{y} \in \mathbb{Z}_p^n$, we have

$$\lim_{j\to+\infty}\int_{\mathbb{Q}_p^n\setminus f^{-1}(0)} \varphi(x)\,|f(x)|_p^{s}\left(p^{nj}\int_{\|y-x\|_p\le p^{-j}} |f(y)|_p^{\beta}\,d^n y\right)d^n x$$

$$= \int_{\mathbb{Q}_p^n\setminus f^{-1}(0)} \varphi(x)\,|f(x)|_p^{s+\beta}\,d^n x,$$

for j big enough and $\mathrm{Re}(s) > 0$. The result follows by using the meromorphic continuation of $|f(x)|_p^{s}$ to the whole complex plane. ∎

Remark 135 Notice that if $-\beta$ is not a pole of $|f|_p^s$, then $T_\beta = |f|_p^{-\beta}$ and $E_\beta = \mathcal{F}^{-1}T_\beta$.

From this observation, by applying Theorem 134 and Corollary 131, we get the following result.

Corollary 136 *Assume that $0 < \beta < \lambda\,(f)$. Then*

$$u\,(x) = \left(\mathcal{F}^{-1}\,|f|_p^{-\beta} * \phi\right)(x) = \int_{\mathbb{Q}_p^n \smallsetminus f^{-1}(0)} \frac{\chi_p\,(-x \cdot \xi)\,(\mathcal{F}\phi)\,(\xi)}{|f\,(\xi)|_p^\beta}d^n\xi$$

is a solution of $f\,(\partial, \beta)\,u = \phi$ in $\mathcal{D}'$, for any $\phi \in \mathcal{D}$.

Set $Dom(f\,(\partial, \beta)) := \left\{\phi \in L^2; |f|_p^\beta\,\mathcal{F}\phi \in L^2\right\}$. Then

$$f\,(\partial, \beta) : Dom(f\,(\partial, \beta)) \to L^2$$

is a well-defined linear operator.

Theorem 137 *If $0 < 2\beta < \lambda\,(f)$, then the following assertions hold:*

(i)   the mapping $v \to E_\beta * v$ is continuous operator from L^2 into L^2;
(ii) $f\,(\partial, \beta)\,u = v$, with $v \in L^2$, has a solution in L^2 given by

$$u\,(x) = \int_{\mathbb{Q}_p^n \smallsetminus f^{-1}(0)} \frac{\chi_p\,(-x \cdot \xi)\,(\mathcal{F}v)\,(\xi)}{|f\,(\xi)|_p^\beta}d^n\xi.$$

Furthermore, if $v \in L^1 \cap L^2$, then u is a continuous function. In this case u is the unique continuous solution of the equation.

Proof

(i) By Corollary 136,

$$\left(E_\beta * v\right)(x) = \int_{\mathbb{Q}_p^n \smallsetminus f^{-1}(0)} \frac{\chi_p\,(-x \cdot \xi)\,(\mathcal{F}v)\,(\xi)}{|f\,(\xi)|_p^\beta}d^n\xi, v \in \mathcal{D}.$$

By using the density of $\mathcal{D}\,(\mathbb{Q}_p^n)$ in L^2 and the Cauchy-Schwarz inequality,

$$\left|E_\beta * v\right| \leq \left(\sqrt{\int_{\mathbb{Q}_p^n \smallsetminus f^{-1}(0)} \frac{d^n\xi}{|f\,(\xi)|_p^{2\beta}}}\right) \|v\|_{L^2} \leq C\,\|v\|_{L^2},$$

we get that $v \to E_\beta * v$ gives rise to a continuous operator from L^2 into L^2.

(ii) Now, since $\mathcal{F}u = |f|_p^{-\beta}\,\mathcal{F}v$, with $v \in L^2$, $\frac{1}{|f|_p^\beta} \in L^1$, by (i), we have $u =$
$\mathcal{F}^{-1}\left(\frac{1}{|f|_p^\beta}\right) * v = E_\beta * v \in L^2$. Finally, if $v \in L^1$, the continuity of u follows from the fact that $\mathcal{F}v \in L^\infty$, by using the dominated convergence theorem. $\blacksquare$

5.5 Fundamental Solutions for Quasielliptic Pseudodifferential Operators

In this section we compute fundamental solutions for quasielliptic pseudodifferential operators, these operators were introduced by Galeano-Peñaloza and Zúñiga-Galindo in [48] as a generalization of the elliptic pseudodifferential operators introduced in [122], see also [52].

Definition 138 Let $f(\xi) \in \mathbb{Z}_p[\xi_1, \ldots, \xi_n]$ be a non-constant polynomial, and let $\omega = (\omega_1, \ldots, \omega_n) \in (\mathbb{N} \smallsetminus \{0\})^n$. We say that f is a quasielliptic polynomial of degree d with respect to ω if the two following conditions hold: (1) $f(\xi) = 0 \Leftrightarrow \xi = 0$, and (2) $f(\lambda^{\omega_1}\xi_1, \ldots, \lambda^{\omega_n}\xi_n) = \lambda^d f(\xi)$, for any $\lambda \in \mathbb{Q}_p^\times$.

For $\omega := (\omega_1, \ldots, \omega_n) \in (\mathbb{N} \smallsetminus \{0\})^n$, we set $|\omega| := \omega_1 + \ldots + \omega_n$.

Proposition 139 Let $f(\xi) \in \mathbb{Z}_p[\xi_1, \ldots, \xi_n]$ be a quasielliptic polynomial of degree d with respect to ω. Then $|f|_p^s$ has a meromorphic continuation to the whole complex plane of the form:

$$\left(|f|_p^s, \phi\right) = \int_{\mathbb{Z}_p^n} \left(\phi(\xi) - \phi(0)\right) |f(\xi)|_p^s\, d^n\xi + \phi(0) \frac{L(p^{-s})}{1 - p^{-|\omega|-ds}}$$

$$+ \int_{\mathbb{Q}_p^n \smallsetminus \mathbb{Z}_p^n} \phi(\xi) |f(\xi)|_p^s\, d^n\xi,$$

where $L(p^{-s})$ is a polynomial in p^{-s} with rational coefficients such that $L\left(p^{\frac{|\omega|}{d}}\right) \neq 0$.

Proof The result follows from the meromorphic continuation of the integral

$$Z(s, f) := Z(s) = \int_{\mathbb{Z}_p^n \smallsetminus \{0\}} |f(\xi)|_p^s\, d^n\xi.$$

More precisely, from the following fact:

Claim

$$Z(s) = \frac{L(p^{-s})}{1 - p^{-|\omega|-ds}}, \tag{5.5}$$

where $L\left(p^{-s}\right)$ is a polynomial in p^{-s} with rational coefficients. In addition, $Z(s)$ has a pole satisfying $\mathrm{Re}\,(s) = \frac{-|\omega|}{d}$.

The proof of this Claim is as follows. Set

$$A := \left\{(\xi_1, \ldots, \xi_n) \in \mathbb{Z}_p^n; ord(\xi_i) \geq w_i \text{ for } i = 1, \ldots, n\right\},$$

and $A^c := \mathbb{Z}_p^n \smallsetminus A$. Then

$$Z(s) = \int_A |f(\xi)|_p^s\, d\xi + \int_{A^c} |f(\xi)|_p^s\, d\xi = p^{-|\omega|-ds} Z(s) + \int_{A^c} |f(\xi)|_p^s\, d\xi,$$

i.e.

$$Z(s) = \frac{1}{1 - p^{-|\omega|-ds}} \int_{A^c} |f(\xi)|_p^s\, d^n\xi.$$

Let I be a proper subset of $\{1, \ldots, n\}$. Set $\boldsymbol{m} = (m_1, \ldots, m_n) \in \mathbb{N}^n$ such that $0 \leq m_i \leq w_i - 1 \Leftrightarrow i \notin I$. Define

$$A\,(I, \boldsymbol{m}) := \left\{(\xi_1, \ldots, \xi_n) \in \mathbb{Z}_p^n; ord(\xi_i) \geq w_i \Leftrightarrow i \in I \text{ and } ord(\xi_i) = m_i \Leftrightarrow i \notin I\right\}.$$

Then $A\,(I, \boldsymbol{m})$ is an open and compact subset of $\mathbb{Z}_p^n$, and A^c is a open and compact subset of $\mathbb{Z}_p^n$ because it is a finite union of subsets of the form $A\,(I, \boldsymbol{m})$. We now show that

$$J_{I, \boldsymbol{m}}\,(s) := \int_{A(I, \boldsymbol{m})} |f(\xi)|_p^s\, d\xi = L_{I, \boldsymbol{m}}\,(p^{-s}),$$

where $L_{I, \boldsymbol{m}}\,(p^{-s})$ is a polynomial in p^{-s} with rational coefficients. Since the origin does not belong to $A\,(I, \boldsymbol{m})$, there exists a finite covering of $A\,(I, \boldsymbol{m})$ by balls $B_i := \tilde{\xi}_i + \left(p^M \mathbb{Z}_p\right)^n$, with $M := M(I, m)$ and $\tilde{\xi}_i \in A\,(I, \boldsymbol{m})$, such that $|f|_p^s\, |_{B_i} = \left|f\left(\tilde{\xi}_i\right)\right|_p^s$ for any i, cf. Lemma 26 and Remark 27 in Chap. 2. Therefore $J_{I, \boldsymbol{m}}\,(s)$ is a polynomial in p^{-s} with rational coefficients, and $Z(s)$ is a rational function of the form

$$Z(s) = \frac{1}{1 - p^{-|\omega|-ds}} \left\{\sum_{(I, \boldsymbol{m})} p^{-(M(I, \boldsymbol{m})n} \left|f\left(\tilde{\xi}_{i(I, \boldsymbol{m})}\right)\right|_p^s\right\}. \tag{5.6}$$

Finally, (5.6) implies that $s = \frac{-|\omega|}{d}$ is a pole of $Z(s)$.   ∎

Definition 140 Let $f(\xi) \in \mathbb{Z}_p[\xi_1, \ldots, \xi_n]$ be a quasielliptic polynomial of degree d with respect to ω. A *quasielliptic pseudodifferential operator* is an extension of an operator of the form

$$\left(f\left(\partial,\beta\right)\phi\right)(x)=\mathcal{F}_{\xi\to x}^{-1}\left(\left|f\left(\xi\right)\right|_{p}^{\beta}\mathcal{F}_{x\to\xi}\phi\right),$$

for $\phi\in\mathcal{D}$.

The family of *quasielliptic pseudodifferential operators includes the Vladimirov operator, the Taibleson operator, cf. Lemma 24 in Sect. 2.3, and the elliptic operators, see Definition 23 in Sect. 2.3.*

Corollary 141 *Let $f\left(\partial,\beta\right)$ be a quasielliptic pseudodifferential operator with $0<2\beta<\frac{|\omega|}{d}$. Then $u\left(x\right)=\int_{\mathbb{Q}_{p}^{n}\smallsetminus\{0\}}\frac{\chi_{p}(-x\cdot\xi)(\mathcal{F}v)(\xi)}{|f(\xi)|_{p}^{\beta}}d^{n}\xi$ is the unique solution in $L^{2}$of $f\left(\partial,\beta\right)u=v$, for $v\in L^{2}$.*

Proof The result follows from Theorem 137 and Proposition 139. ∎

Theorem 142 *Let $f\left(\partial,\beta\right)$ be a quasielliptic pseudodifferential operator, and let E_{β} be fundamental solution for $f\left(\partial,\beta\right)u=\phi$, with $\phi\in\mathcal{D}$.*

(1) If $\beta\neq\frac{|\omega|}{d}$, then

$$\left(E_{\beta},\phi\right)=\int_{\mathbb{Z}_{p}^{n}}\frac{\left(\mathcal{F}^{-1}\phi\right)(x)-\left(\mathcal{F}^{-1}\phi\right)(0)}{|f\left(x\right)|_{p}^{\beta}}d^{n}x+\left(\mathcal{F}^{-1}\phi\right)(0)\frac{L\left(p^{\beta}\right)}{1-p^{-|\omega|+d\beta}}$$

$$+\int_{\mathbb{Q}_{p}^{n}\smallsetminus\mathbb{Z}_{p}^{n}}\frac{\left(\mathcal{F}^{-1}\phi\right)(x)}{|f\left(x\right)|_{p}^{\beta}}d^{n}x,\qquad(5.7)$$

for $\phi\in\mathcal{D}$, where L is the polynomial in p^{-s} defined in Proposition 139.
(2) If $\beta=\frac{|\omega|}{d}$, then

$$\left(E_{\frac{|\omega|}{d}},\phi\right)=\int_{\mathbb{Z}_{p}^{n}}\frac{\left(\mathcal{F}^{-1}\phi\right)(x)-\left(\mathcal{F}^{-1}\phi\right)(0)}{|f\left(x\right)|_{p}^{\frac{|\omega|}{d}}}d^{n}x$$

$$+\left\{\frac{1}{2}L\left(p^{\frac{|\omega|}{d}}\right)-\frac{1}{d}L'\left(p^{\frac{|\omega|}{d}}\right)\right\}\left(\mathcal{F}^{-1}\phi\right)(0)+\int_{\mathbb{Q}_{p}^{n}\smallsetminus\mathbb{Z}_{p}^{n}}\frac{\left(\mathcal{F}^{-1}\phi\right)(x)}{|f\left(x\right)|_{p}^{\frac{|\omega|}{d}}}d^{n}x,$$

$$(5.8)$$

for $\phi\in\mathcal{D}$, where L is the polynomial in p^{-s} defined in Proposition 139, and L' is its derivative.

Proof By Proposition 139, the poles of $|f|_{p}^{s}$ have the form $-\frac{|\omega|}{d}+\frac{2\pi i\mathbb{Z}}{d\ln p}$. If $\beta\neq\frac{|\omega|}{d}$, then $T_{\beta}=|f|_{p}^{-\beta}$. An explicit expression for T_{β} is obtained from the formula for the meromorphic continuation of $|f|_{p}^{s}$ given in Proposition 139, then (5.7) follows from $E_{\beta}=\mathcal{F}^{-1}T_{\beta}$.

If $\beta = \frac{|\omega|}{d}$, by the proof of Theorem 134 we know that T_β is the constant term of the Laurent expansion of the meromorphic continuation of $|f|_p^s$ at $s = -\frac{|\omega|}{d}$. To carry out this calculation, we note that $\phi(0) \frac{L(p^{-s})}{1-p^{-|\omega|-ds}}$ equals

$$
\phi(0) \left(\frac{1}{d(\ln p)\left(s + \frac{|\omega|}{d}\right)} + \left\{ \frac{1}{2} L\left(p^{\frac{|\omega|}{d}}\right) - \frac{1}{d} L'\left(p^{\frac{|\omega|}{d}}\right) \right\} + O\left(s + \frac{|\omega|}{d}\right) \right),
$$

where $O\left(\left(s + \frac{|\omega|}{d}\right)\right)$ denotes a holomorphic function. Thus, by Proposition 139, the constant term of the Laurent expansion of $|f|_p^s$ at $s = -\frac{|\omega|}{d}$ is

$$
\left(T_{\frac{|\omega|}{d}}, \phi \right) = \int_{\mathbb{Z}_p^n} \frac{(\phi(x) - \phi(0))}{|f(x)|_p^{\frac{|\omega|}{d}}} d^n x + \left\{ \frac{1}{2} L\left(p^{\frac{|\omega|}{d}}\right) - \frac{1}{d} L'\left(p^{\frac{|\omega|}{d}}\right) \right\} \phi(0)
$$

$$
+ \int_{\mathbb{Q}_p^n \smallsetminus \mathbb{Z}_p^n} \frac{\phi(x)}{|f(x)|_p^{\frac{|\omega|}{d}}} d^n x.
$$

Finally, (5.8) follows from $E_{\frac{|\omega|}{d}} = \mathcal{F}^{-1} T_{\frac{|\omega|}{d}}$.   ∎

Take $n = 1$ and $f(\xi) = \xi$. Then $f(\partial, \beta) = D^\beta$ is the Vladimirov operator, see [111, IX]. We also set $f_\beta(x) := \frac{|x|_p^{\beta-1}}{\Gamma_1(\beta)}$, with $\Gamma_1(\beta) = \frac{1-p^{\beta-1}}{1-p^{-\beta}}$, for the Riesz kernel.

Corollary 143 *With the above notation,*

$$
E_\beta(x) = \begin{cases} f_\beta(x) & \text{if } \beta \neq 1 \\[2mm] \frac{1-p}{p \ln p} \ln |x|_p & \text{if } \beta = 1 \end{cases}
$$

is a fundamental solution for D^β.

Proof The case $\beta \neq 1$ follows directly from (5.7), by using that $\omega = d = 1$, $L(p^\beta) = 1 - p^{-1}$ and that $\mathcal{F}(f_\beta(x)) = |x|_p^{-\beta}$ as distributions.

We now consider the case $\beta = 1$. From (5.8) one gets

$$
(E_1, \phi) = \int_{\mathbb{Z}_p} \frac{(\mathcal{F}^{-1}\phi)(x) - (\mathcal{F}^{-1}\phi)(0)}{|x|_p} d^n x + \int_{\mathbb{Q}_p \smallsetminus \mathbb{Z}_p} \frac{(\mathcal{F}^{-1}\phi)(x)}{|x|_p} d^n x
$$

$$
+ \frac{(1-p^{-1})}{2} (\mathcal{F}^{-1}\phi)(0) = \left(\mathcal{P}|x|_p^{-1}, \mathcal{F}^{-1}\phi \right) + \frac{(1-p^{-1})}{2} (\mathcal{F}^{-1}\phi)(0).
$$

We now recall that

$$\mathcal{F}^{-1}\left(P\,|x|_p^{-1}\right) = \frac{1-p}{p\ln p}\ln|x|_p - p^{-1} \text{ in } \mathcal{D}'\left(\mathbb{Q}_p\right),$$

see e.g. [80, p. 39], and since $\left(\mathcal{F}^{-1}\phi\right)(0) = \left(\delta, \mathcal{F}^{-1}\phi\right) = \left(\mathcal{F}^{-1}\delta, \phi\right) = (1, \phi)$, then

$$E_1 = \frac{1-p}{p\ln p}\ln|x|_p + \left(\frac{p-3}{2p}\right) \text{ in } \mathcal{D}'\left(\mathbb{Q}_p\right). \tag{5.9}$$

Now, since

$$\left(\boldsymbol{D}^\beta\phi\right)(x) = \frac{1}{\Gamma_1(\beta)}\int_{\mathbb{Q}_p}|x|_p^{-\beta-1}\left[\phi(x-y) - \phi(0)\right]dy,$$

we have that if E_β is a fundamental solution of $\boldsymbol{D}^\beta$, then $E_\beta + c$, with $c \in \mathbb{Q}_p$, is also a fundamental solution for $\boldsymbol{D}^\beta$. The second case follows from (5.9) by replacing E_1 by $E_1 - \frac{p-3}{2p}$. ∎

5.6 Schrödinger-Type Pseudodifferential Equations

In this section we use some basic geometric aspects of the p-adic manifolds. For further details the reader may consult [57] or [104]. At any rate, we use these notions in a very particular case in which, we think, the reader may follow the presentation easily.

Set $h(\xi, \tau) = \tau - h_0(\xi) \in \mathbb{Z}_p[\xi_1, \dots, \xi_n, \tau]$ with $h_0(0) = 0$. In addition, we assume that $\nabla h_0(\xi) = 0 \Leftrightarrow \xi = 0$. We set

$$\left(|h|_p^s, \phi\right) = \int_{\mathbb{Q}_p^{n+1}\setminus h^{-1}(0)} \phi(\xi, \tau)\,|\tau - h_0(\xi)|_p^s\,d\tau d^n\xi, \text{ for } \mathrm{Re}(s) > 0. \tag{5.10}$$

Notice that the hypersurface $V_h(\mathbb{Q}_p) := V_h = \left\{(\tau, \xi) \in \mathbb{Q}_p^{n+1} : h(\tau, \xi) = 0\right\}$ does not have singular points, i.e. the system of equations

$$h(\xi, \tau) = \nabla h(\xi, \tau) = 0$$

has no solutions in $\mathbb{Q}_p^{n+1}$. Technically speaking, V_h is a closed submanifold of codimension one of $\mathbb{Q}_p^{n+1}$.

Set $V_{h,\lambda} = \{(\xi, \tau) \in \mathbb{Q}_p^{n+1} : \tau - h_0(\xi) = \lambda\}$, for $\lambda \in \mathbb{Q}_p$ fixed. Then $V_{h,\lambda}$ is a closed submanifold of codimension one for every λ. For $\phi(\xi, \tau) \in \mathcal{D}(\mathbb{Q}_p^{n+1})$, we take

$$F_\phi(\lambda) := \int_{V_{h,\lambda}} \phi(\tau, \xi) |\gamma_{GL}|, \qquad (5.11)$$

where γ_{GL} is a Gel'fand-Leray form along $V_{h,\lambda}$. On other hand, by *using integration on the fibers*, see Chap. 1, formula (1.3) or [57, Section 7.6], one gets from (5.10)–(5.11):

$$\left(|h|_p^s, \phi\right) = \int_{\mathbb{Q}_p} |\lambda|_p^s F_\phi(\lambda) \, d\lambda, \text{ for } \mathrm{Re}(s) > 0. \qquad (5.12)$$

Now by using the meromorphic continuation of the distribution $|\lambda|_p^s$, the fact that $F_\phi(\lambda) \in \mathcal{D}(\mathbb{Q}_p)$, and (5.11), we get a meromorphic continuation for $|h|_p^s$.

The distribution $|h|_p^s$, $\mathrm{Re}(s) > 0$, has the following meromorphic continuation to the whole complex plane:

$$\left(|h|_p^s, \phi\right) = \int_{\mathbb{Z}_p} |\lambda|_p^s \{F_\phi(\lambda) - F_\phi(0)\} \, d\lambda + \left(\frac{1 - p^{-1}}{1 - p^{-1-s}}\right) F_\phi(0)$$

$$+ \int_{\mathbb{Q}_p \smallsetminus \mathbb{Z}_p} |\lambda|_p^s F_\phi(\lambda) \, d\lambda.$$

We now observe that

$$\mathcal{F}_{\lambda \to z}^{-1}\left(F_\phi(\lambda)\right) = \int_{\mathbb{Q}_p} \chi_p(-\lambda z) \left\{ \int_{V_{h,\lambda}} \phi(\tau, \xi) |\gamma_{GL}| \right\} d\lambda$$

$$= \int_{\mathbb{Q}_p^{n+1}} \chi_p(-z\tau + zh_0(\xi)) \phi(\tau, \xi) \, d\tau d^n\xi d\lambda, \qquad (5.13)$$

for further details the reader may consult [57, Theorem 8.3.1]. Hence

$$\mathcal{F}_{\lambda \to z}^{-1}\left(F_\phi(\lambda)\right)\big|_{z=0} = \int_{\mathbb{Q}_p^{n+1}} \phi(\tau, \xi) \, d\tau d^n\xi. \qquad (5.14)$$

Definition 144 An extension of an operator of the form

$$(\boldsymbol{h}(\partial, \beta)\phi)(x, t) = \mathcal{F}_{\substack{\tau \to t \\ \xi \to x}}^{-1}\left(|\tau - h_0(\xi)|_p^\beta \, \mathcal{F}_{\substack{t \to \tau \\ x \to \xi}}\phi\right), \text{ for } \phi \in \mathcal{D}(\mathbb{Q}_p^{n+1})$$

is called a Schrödinger-type pseudodifferential operator.

These operators were introduced by Kochubei, see [80] and references therein, and also [120].

By Proposition 5.6 and Theorem 137 we get the following result:

Corollary 145 *If* $0 < 2\beta < 1$, *then* $u(x,t) = \int_{\mathbb{Q}_p^{n+1}} \frac{\chi_p(-x\cdot\xi - t\tau)(\mathcal{F}v)(\xi)}{|\tau - h_0(\xi)|_p^\beta} d^n\xi d\tau$ *is the is the unique solution in* L^2 *of* $\mathbf{h}(\partial, \beta)u = v$, *for* $v \in L^2$.

From Proposition 5.6 by using the technique presented in the proof of Theorem 134, and the formulas (5.13)–(5.14), we get the following result.

Theorem 146 *With the above notation. Let* E_β *be fundamental solution for* $\mathbf{h}(\partial, \beta)u = \phi$, *with* $\phi \in \mathcal{D}$.

(1) If $\beta \neq 1$, *then*

$$
(E_\beta, \phi) = \int_{\mathbb{Z}_p} \int_{\mathbb{Q}_p^{n+1}} \frac{\{\chi_p(-\lambda\tau + \lambda h_0(\xi)) - 1\}\phi(\tau, \xi)}{|\lambda|_p^\beta} d\tau d^n\xi d\lambda
$$

$$
+ \left(\frac{1 - p^{-1}}{1 - p^{-1+\beta}}\right) \int_{\mathbb{Q}_p^{n+1}} \phi(\tau, \xi) d\tau d^n\xi
$$

$$
+ \int_{\mathbb{Q}_p \setminus \mathbb{Z}_p} \int_{\mathbb{Q}_p^{n+1}} \frac{\chi_p(-\lambda\tau + \lambda h_0(\xi))\phi(\tau, \xi)}{|\lambda|_p^\beta} d\tau d^n\xi d\lambda.
$$

(2) If $\beta = 1$, *then*

$$
(E_1, \phi) = \int_{\mathbb{Z}_p} \int_{\mathbb{Q}_p^{n+1}} \frac{\{\chi_p(-\lambda\tau + \lambda h_0(\xi)) - 1\}\phi(\tau, \xi)}{|\lambda|_p} d\tau d^n\xi d\lambda
$$

$$
+ \left(\frac{1 - p^{-1}}{2}\right) \int_{\mathbb{Q}_p^{n+1}} \phi(\tau, \xi) d\tau d^n\xi
$$

$$
+ \int_{\mathbb{Q}_p \setminus \mathbb{Z}_p} \int_{\mathbb{Q}_p^{n+1}} \frac{\chi_p(-\lambda\tau + \lambda h_0(\xi))\phi(\tau, \xi)}{|\lambda|_p} d\tau d^n\xi d\lambda.
$$

Corollary 147 *Take* $\beta > 1$ *and* $v \in \mathcal{D}(\mathbb{Q}_p^{n+1})$ *satisfying* $\int_{\mathbb{Q}_p^{n+1}} v(t, x) dt d^n x = 0$. *Then*

$$
u_1(x,t) = \int_{\mathbb{Q}_p} \int_{\mathbb{Q}_p^{n+1}} \frac{\{\chi_p(-\lambda\tau + \lambda h_0(\xi)) - 1\}v(t - \tau, x - \xi)}{|\lambda|_p^\beta} d\tau d^n\xi d\lambda \in \mathcal{D}(\mathbb{Q}_p^{n+1})
$$

is a solution of $\mathbf{h}(\partial, \beta)u_1 = v$.

5.6.1 The Homogeneous Equation

Theorem 148 *Consider*

$$\boldsymbol{h}\left(\partial, \beta\right) : \left\{v \in L^2\left(\mathbb{Q}_p^{n+1}\right) ; \hat{v} \, |\tau - h_0\left(\xi\right)|_p^\beta \in L^2\left(\mathbb{Q}_p^{n+1}\right)\right\} \to L^2\left(\mathbb{Q}_p^{n+1}\right).$$

Then, the initial value problem:

$$\begin{cases} \left(\boldsymbol{h}\left(\partial, \beta\right) u_0\right)(x, t) = 0, \, x \in \mathbb{Q}_p^n, \, t \in \mathbb{Q}_p \\[2mm] u_0(x, 0) = \phi \in \mathcal{D}\left(\mathbb{Q}_p^n\right), \end{cases} \tag{5.15}$$

has a unique solution in $\mathcal{D}'\left(\mathbb{Q}_p^{n+1}\right)$ given by

$$u_0\left(x, t\right) = \int\limits_{\mathbb{Q}_p^n} \chi_p\left(-t h_0\left(\xi\right) - \xi \cdot x\right) (\mathcal{F}\phi)\left(\xi\right) d^n \xi, \tag{5.16}$$

this function is locally constant in (x, t) and square-integrable in x for any $t \in \mathbb{Q}_p$. If the datum ϕ in (5.15) can be taken in $L^2\left(\mathbb{Q}_p^n\right)$, then (5.16) is the unique solution in $\mathcal{D}'\left(\mathbb{Q}_p^{n+1}\right)$ of (5.15).

Proof We first established the theorem in the case $\phi \in \mathcal{D}\left(\mathbb{Q}_p^n\right)$. Any non-trivial solution $u_0 \in L^2$, i.e. u_0 is not zero almost everywhere, of $\boldsymbol{h}\left(\partial, \beta\right) u_0 = 0$ necessarily satisfies ess supp $\mathcal{F}u_0 \subset V_h$, where *ess supp*$(\cdot)$ denotes the essential support of a measurable function. Let $1_{V_h}\left(\xi, \tau\right)$ denote the characteristic function of V_h, and set $(\mathcal{F}u_0)\left(\xi, \tau\right) := (\mathcal{F}\phi)\left(\xi\right) \cdot 1_{V_h}\left(\xi, \tau\right)$, for $\phi \in \mathcal{D}\left(\mathbb{Q}_p^n\right)$. From the fact that $1_{V_h}\left(\xi, \tau\right) \cdot 1_{\text{supp}\mathcal{F}u_0}\left(\xi\right)$ has compact support in $\mathbb{Q}_p^{n+1}$, it follows that $(\mathcal{F}u_0)\left(\xi, \tau\right)$ is a bounded function with compact essential support contained in V_h that satisfies

$$|\tau - h_0\left(\xi\right)|_p^\beta \left(\mathcal{F}u_0\right)\left(\xi, \tau\right) = 0. \tag{5.17}$$

Consider for $\tau \in \mathbb{Q}_p \smallsetminus \{0\}$ fixed, the distribution on $\mathcal{D}\left(\mathbb{Q}_p^n\right)$:

$$\left((\mathcal{F}\phi)\left(\xi\right) \cdot 1_{V_h}\left(\xi, \tau\right), \theta\left(\xi\right)\right) = \int\limits_{\{h_0(\xi) = \tau\}} (\mathcal{F}\phi)\left(\xi\right) \theta\left(\xi\right) |v_{GL}|,$$

where v_{GL} is a Gel'fand-Leray differential form on $\{h_0\left(\xi\right) = \tau\}$, we note $v_{GL} \neq d\xi_1 \wedge \cdots d\xi_n$ because τ is fixed, the existence of the form v_{GL} is a consequence of the fact $\{h_0\left(\xi\right) = \tau\}$, τ fixed in $\mathbb{Q}_p \smallsetminus \{0\}$, is a non-singular hypersurface because $\nabla h_0\left(\xi\right) = 0 \Leftrightarrow \xi = 0$. Now, since $(\mathcal{F}\phi)\left(\xi\right) \cdot 1_{V_h}\left(\xi, \tau\right)$, $\tau \in \mathbb{Q}_p \smallsetminus \{0\}$ fixed, is a distribution with compact support, we have

$$\mathcal{F}^{-1}_{\xi \to x} \left((\mathcal{F}\phi) \, (\xi) \cdot 1_{V_h} \, (\xi, \tau) \right) = \int\limits_{\{h_0(\xi) = \tau\}} \chi_p \, (-\xi \cdot x) \, (\mathcal{F}\phi) \, (\xi) \, |v_{GL}|$$

in $\mathcal{D}' \left(\mathbb{Q}_p^n \right)$. We now notice that the function

$$\tau \to \int\limits_{\{h_0(\xi) = \tau\}} \chi_p \, (-\xi \cdot x) \, (\mathcal{F}\phi) \, (\xi) \, |v_{GL}|$$

has compact support because $h_0 \, (\text{supp} \mathcal{F}\phi)$ is compact in $\mathbb{Q}_p$. Then

$$u_0 \, (x, t) := \mathcal{F}^{-1}_{\tau \to t} \mathcal{F}^{-1}_{\xi \to x} \left((\mathcal{F}\phi) \, (\xi) \cdot 1_{V_h} \, (\xi, \tau) \right)$$

$$= \int\limits_{\mathbb{Q}_p \smallsetminus \{0\}} \int\limits_{\{h_0(\xi) = \tau\}} \chi_p \, (-\tau t - \xi \cdot x) \, (\mathcal{F}\phi) \, (\xi) \, |v_{GL}| \, d\tau$$

in $\mathcal{D}' \left(\mathbb{Q}_p^{n+1} \right)$. We now recall that $\{\xi \in \mathbb{Q}_p^n; h_0 \, (\xi) = \lambda\}$ is a non-singular hypersurface for $\lambda \neq 0$, then

$$\int\limits_{\mathbb{Q}_p^n} \theta \, (\xi) \, d^n \xi = \int\limits_{\mathbb{Q}_p \smallsetminus \{0\}} \left\{ \int\limits_{\{h_0(\xi) = \lambda\}} \theta \, (\xi) \, |v_{GL}| \right\} d\lambda$$

see [57, Section 7.6], hence

$$u_0 \, (x, t) = \int\limits_{\mathbb{Q}_p^n} \chi_p \, (-t h_0 \, (\xi) - \xi \cdot x) \, (\mathcal{F}\phi) \, (\xi) \, d^n \xi \text{ in } \mathcal{D}' \left(\mathbb{Q}_p^{n+1} \right).$$

We note that $u_0 \, (x, t)$ is a locally constant function since it is the inverse Fourier transform of a distribution with compact support. Furthermore, since $u_0 \, (x, t) = \mathcal{F}^{-1}_{\xi \to x} \left(\chi_p \, (t h_0 \, (\xi)) \, (\mathcal{F}\phi) \, (\xi) \right)$, $u_0 \, (\cdot, t) \in L^2$ for any t. The uniqueness of the solution follows from the fact that any non-trivial solution has the form (5.16) for some $\phi \in \mathcal{D}$.

We now take $\phi \in L^2 \left(\mathbb{Q}_p^n \right)$. Notice that (5.17) is still valid in this case. To establish the formula (5.16) we take replace $(\mathcal{F}\phi)$ by $\Delta_k \, (\mathcal{F}\phi)$, where Δ_k is the characteristic function of the ball B_k^n, an set $\left(\mathcal{F} u_0^{(k)} \right) (\xi, \tau) := \Delta_k \, (\xi) \, (\mathcal{F}\phi) \, (\xi) \cdot 1_{V_h} \, (\xi, \tau)$. Then

$$u_0^{(k)} \, (x, t) = \int\limits_{\|\xi\|_p \leq p^k} \chi_p \, (-t h_0 \, (\xi) - \xi \cdot x) \, (\mathcal{F}\phi) \, (\xi) \, d^n \xi \text{ in } \mathcal{D}' \left(\mathbb{Q}_p^{n+1} \right).$$

Then $u_0^{(k)} \, (\cdot, t) \xrightarrow{L^2} u_0 \, (\cdot, t)$.   ∎

Theorem 149

(i) *The operator $v \to T_t v$, from L^2 into L^2, where $t \in \mathbb{Q}_p$ and*

$$(T_t v)(x) := \int_{\mathbb{Q}_p^n} \chi_p(-t h_0(\xi) - \xi \cdot x)(\mathcal{F} v)(\xi) \, d^n \xi$$

is linear continuous, and satisfies: (1) $T_0 = I$ is the identity operator on L^2; (2) $T_{t+s} = T_t T_s$ for any $t, s \in \mathbb{Q}_p$; (3) $T_t v \to v$ as $t \to 0$ in L^2; (4) $\mathbf{D}_t T_t v = \mathbf{h}_0(\partial, \beta) T_t v$, for any $v \in L^2$, $t \in \mathbb{Q}_p$, where $\mathbf{D}_t$ is the Vladimirov operator and $(\mathbf{h}_0(\partial, \beta)\phi)(x) = \mathcal{F}_{\xi \to x}^{-1}\left(|h_0(\xi)|_p^{\beta}\mathcal{F}_{x \to \xi}\phi\right)$, for $\phi \in \mathcal{D}(\mathbb{Q}_p^n)$.

(ii) *The function $u_0(x, t) = (T_t v)(x)$ is the unique solution of the following initial value problem:*

$$\begin{cases} (\mathbf{h}(\partial, \beta) u_0)(x, t) = 0 \\ \\ u_0(x, 0) = v, \qquad v \in L^2. \end{cases} \tag{5.18}$$

Proof

(i) Notice that $(T_t v)(x) = \mathcal{F}_{\xi \to x}^{-1}\left(\chi_p(t h_0(\xi)) \mathcal{F}_{x \to \xi} v\right)$. From this identity one gets (1), (2) and (3). (4) By using the fact that $\mathcal{D}$ is dense in L^2, it is sufficient to established the announced formula in the case of test functions. Take $\phi \in \mathcal{D}$, since

$$(T_t \phi)(x) = \mathcal{F}_{\tau \to t}^{-1}\left(\left\{\int_{\tau - h_0(\xi) = 0} \chi_p(-\xi \cdot x)(\mathcal{F}\phi)(\xi) |v_{GL}|\right\}\right),$$

in $\mathcal{D}'(\mathbb{Q}_p)$, and the Vladimirov operator extends to distributions having Fourier transform with compact support, we have

$$\mathbf{D}_t (T_t \phi)(x) = \int_{\mathbb{Q}_p} \chi_p(-t\tau) |\tau|_p \left\{\int_{\tau - h_0(\xi) = 0} \chi_p(-\xi \cdot x)(\mathcal{F}\phi)(\xi) |v_{GL}|\right\} d\tau$$

$$= \int_{\mathbb{Q}_p^n} |h_0(\xi)|_p \chi_p(-t h_0(\xi) - \xi \cdot x)(\mathcal{F}\phi)(\xi) \, d^n \xi.$$

On the other hand, since $(T_t \phi)(x) = \mathcal{F}_{\xi \to x}^{-1}\left(\chi_p(-t h_0(\xi)) \mathcal{F}_{x \to \xi}\phi\right)$, we have

$$(\mathbf{h}_0(\partial, \beta) T_t)\phi(x) = \int_{\mathbb{Q}_p^n} |h_0(\xi)|_p \chi_p(-t h_0(\xi) - \xi \cdot x)(\mathcal{F}\phi)(\xi) \, d^n \xi,$$

therefore $\mathbf{D}_t (T_t \phi)(x) = (\mathbf{h}_0(\partial, \beta) T_t)\phi(x)$.

(ii) It follows from Theorem 148. ∎

5.6.2   The Inhomogeneous Equation

Theorem 150 *Let $\beta > 1$ and $v \in \mathcal{D}\left(\mathbb{Q}_p^{n+1}\right)$ satisfying $\int_{\mathbb{Q}_p^{n+1}} v(t, x)\, dt d^n x = 0$, and let $u_0(x) \in \mathcal{D}\left(\mathbb{Q}_p^n\right)$. Consider the following initial value problem:*

$$
\begin{cases}
(\boldsymbol{h}(\partial, \beta) u)(x, t) = v(x, t), \ x \in \mathbb{Q}_p^n, \ t \in \mathbb{Q}_p \\[2mm]
u(x, 0) = u_0(x).
\end{cases}
\tag{5.19}
$$

Set

$$
u_0'(x) := \int\limits_{\mathbb{Q}_p \mathbb{Q}_p^{n+1}} \int \frac{\left\{\chi_p\left(-\lambda \tau + \lambda h_0(\xi)\right) - 1\right\} v(-\tau, x - \xi)}{|\lambda|_p^{\beta}}\, d\tau d^n \xi d\lambda
$$

Then $u_0'(x) \in \mathcal{D}\left(\mathbb{Q}_p^n\right)$ and

$$
u(x, t) = \int\limits_{\mathbb{Q}_p^n} \chi_p\left(-t h_0(\xi) - \xi \cdot x\right)\left(\mathcal{F}\left(u_0 - u_0'\right)\right)(\xi)\, d^n \xi
$$

$$
+ \int\limits_{\mathbb{Q}_p \mathbb{Q}_p^{n+1}} \int \frac{\left\{\chi_p\left(-\lambda \tau + \lambda h_0(\xi)\right) - 1\right\} v(t - \tau, x - \xi)}{|\lambda|_p^{\beta}}\, d\tau d^n \xi d\lambda
$$

is the unique solution in $\mathcal{D}'\left(\mathbb{Q}_p^{n+1}\right)$ of (5.19).

Proof From Theorem 148 and Corollary 147 imply that $u(x, t)$ is a solution of (5.19). The uniqueness of the solution is a consequence of Theorem 148. ∎

Chapter 6
Pseudodifferential Equations of Klein-Gordon Type

6.1 Introduction

In the 1980s I. Volovich proposed that the world geometry in regimes smaller than the Planck scale might be non-Archimedean [112, 113]. This hypothesis conducts naturally to consider models involving geometry and analysis over $\mathbb{Q}_p$. Since then, a big number of articles have appeared exploring these and related themes, see e.g. [36], [107, Chapter 6] and the references therein.

In this chapter, which is based on [119], we introduce a new class of non-Archimedean pseudodifferential equations of Klein-Gordon type. We work on the p-adic Minkowski space which is the quadratic space $\left(\mathbb{Q}_p^4, Q\right)$ where $Q(k) = k_0^2 - k_1^2 - k_2^2 - k_3^2$. Our starting point is a result of Rallis-Schiffmann that asserts the existence of a unique measure on $V_t = \left\{k \in \mathbb{Q}_p^4; Q(k) = t\right\}$ which is invariant under the orthogonal group $O(Q)$ of Q, see Proposition 153 or [92]. By using Gel'fand-Leray differential forms, we reformulate this results in terms of Dirac distributions $\delta\left(Q(k) - t\right)$ invariant under $O(Q)$, see Remark 154 and Lemma 156. We introduce the positive and negative mass shells $V_{m^2}^+$ and $V_{m^2}^-$, here m is the 'mass parameter' which is taken to be a nonzero p-adic number. The restriction of $\delta\left(Q(k) - t\right)$ to $V_{m^2}^\pm$ gives two distributions $\delta_\pm\left(Q(k) - t\right)$ which are invariant under $\mathcal{L}_+^\uparrow$, the 'Lorentz proper group', see Definition 162, and that satisfy $\delta\left(Q(k) - t\right) = \delta_+\left(Q(k) - t\right) + \delta_-\left(Q(k) - t\right)$, see Lemma 163. The p-adic Klein-Gordon type pseudodifferential operators introduced here have the form

$$\left(\Box_{\alpha,m}\varphi\right)(x) = \mathcal{M}_{k \to x}^{-1}\left[\left|Q(k) - m^2\right|_p^\alpha \mathcal{M}_{x \to k}\varphi\right], \quad \alpha > 0, \, m \in \mathbb{Q}_p \smallsetminus \{0\},$$

where $\mathcal{M}$ denotes the Fourier-Minkowski transform. We solve the Cauchy problem for these operators, see Theorem 174.

© Springer International Publishing AG 2016
W.A. Zúñiga-Galindo, *Pseudodifferential Equations Over Non-Archimedean Spaces*, Lecture Notes in Mathematics 2174,
DOI 10.1007/978-3-319-46738-2_6

The equations $(\Box_{\alpha,m}\phi)(t,x) = 0$ have many similar properties to the classical Klein-Gordon equations, see e.g. [31, 32, 103]. These equations admit plane waves as weak solutions, see Lemma 177; the distributions $a\mathcal{M}^{-1}\left[\delta_{+}\left(Q(k) - m^2\right)\right] + b\mathcal{M}^{-1}\left[\delta_{-}\left(Q(k) - m^2\right)\right]$, $a, b \in \mathbb{C}$, are weak solutions of these equations, see Proposition 171. The locally constant functions

$$\phi(t,x) = \int_{U_{Q,m}} \chi_p(x \cdot k) \left\{\chi_p(-t\omega(k))\phi_{+}(k) + \chi_p(t\omega(k))\phi_{-}(k)\right\} d^3k, \qquad (6.1)$$

where $\chi_p(\cdot)$ denotes the standard additive character of $\mathbb{Q}_p$ and $\phi_{\pm}$ are locally constant functions with support in $U_{Q,m}$, are weak solutions of these equations.

At this point, it is relevant to mention that the operators, equations and techniques introduced here are new. In [5] a very general theory for pseudodifferential operators and equations involving symbols that vanish only at the origin was developed. This theory cannot be applied here because our symbols $(\left|Q(k) - m^2\right|_p^{\alpha})$ have infinitely many zeros.

Finally, we want to mention that another type of p-adic hyperbolic equations was introduced by Kochubei in [79].

6.2 Preliminaries

Remark 151 Along this chapter p will denote a prime number different from 2. We recall that any non-zero p-adic number x can be written uniquely as $p^{ord(x)}ac(x)$, where $ac(x) \in \mathbb{Z}_p^{\times}$ is the angular component of x. For $x = x_0 + x_1p + \cdots \in \mathbb{Z}_p^{\times}$, we denote by $\bar{x}$ the digit x_0 considered as an element of $\mathbb{F}_p$, the finite field with p elements.

6.2.1 Fourier Transform on Finite Dimensional Vector Spaces

Let E be a finite dimensional vector space over $\mathbb{Q}_p$ and χ_p a non-trivial additive character of $\mathbb{Q}_p$ as before. Let $[x, y]$ be a symmetric non-degenerate $\mathbb{Q}_p$- bilinear form on $E \times E$. Thus $Q(e) := [e, e]$, $e \in E$ is a *non-degenerate quadratic form* on E. We identify E with its algebraic dual E^* by means of $[\cdot, \cdot]$. We now identify the dual group (i.e. the Pontryagin dual) of $(E, +)$ with E^* by taking $\langle e, e^* \rangle = \chi_p([e, e^*])$ where $[e, e^*]$ is the algebraic duality. The Fourier transform takes the form

$$\hat{\varphi}(y) = \int_E \varphi(x) \chi_p([x, y]) dx \text{ for } \varphi \in L^1(E),$$

where dx is a Haar measure on E.

Let $\mathcal{L}(E)$ be the space of continuous functions φ in $L^1(E)$ whose Fourier transform $\hat{\varphi}$ is in $L^1(E)$. The measure dx can be normalized uniquely in such manner that $\widehat{(\hat{\varphi})}(x) = \varphi(-x)$ for every φ belonging to $\mathcal{L}(E)$. We say that dx is *a self-dual measure relative to* $\chi_p([\cdot, \cdot])$.

For further details about the material presented in this section the reader may consult [115].

6.2.2   The p-Adic Minkowski Space

We take E to be the $\mathbb{Q}_p$-vector space of dimension 4. By fixing a basis we identify E with $\mathbb{Q}_p^4$ considered as a $\mathbb{Q}_p$-vector space. For $x = (x_0, x_1, x_2, x_3) := (x_0, \boldsymbol{x})$ and $y = (y_0, y_1, y_2, y_3) := (y_0, \boldsymbol{y})$ in $\mathbb{Q}_p^4$ we set

$$[x, y] := x_0 y_0 - x_1 y_1 - x_2 y_2 - x_3 y_3 := x_0 y_0 - \boldsymbol{x} \cdot \boldsymbol{y}, \tag{6.2}$$

which is a symmetric non-degenerate bilinear form. *From now on, we use* $[x, y]$ *to mean bilinear form (6.2).* Then $(\mathbb{Q}_p^4, Q)$, with $Q(x) = [x, x]$ is a *quadratic vector space* and Q is a non-degenerate quadratic form on $\mathbb{Q}_p^4$. We will call $(\mathbb{Q}_p^4, Q)$ *the p-adic Minkowski space.*

On $(\mathbb{Q}_p^4, Q)$, the Fourier-Minkowski transform takes the form

$$\mathcal{M}[\varphi](k) = \int_{\mathbb{Q}_p^4} \chi_p([x, k]) \, \varphi(x) \, d^4 x \text{ for } \varphi \in L^1(\mathbb{Q}_p^4), \tag{6.3}$$

where $d^4 x$ is a self-dual measure for $\chi_p([\cdot, \cdot])$, i.e. $\mathcal{M}[\mathcal{M}[\varphi]](x) = \varphi(-x)$ for every φ belonging to $\mathcal{L}(\mathbb{Q}_p^4)$. Notice that $d^4 x$ is equal to a positive multiple of the normalized Haar measure on $\mathbb{Q}_p^4$, i.e. $d^4 x = C d^4 \mu(x)$, with $C > 0$.

Remark 152

(i) We recall that the usual Fourier transform $\mathcal{F}$ on $\mathbb{Q}_p^4$ has the form

$$\mathcal{F}[\varphi][k] = \int_{\mathbb{Q}_p^4} \chi_p(x_0 k_0 + x_1 k_1 + x_2 k_2 + x_3 k_3) \, \varphi(x) \, d^4 \mu(x) \text{ for } \varphi \in L^1(\mathbb{Q}_p^4),$$

where $d^4 \mu(x)$ is the normalized Haar measure of $\mathbb{Q}_p^4$. The connection between $\mathcal{M}$ and $\mathcal{F}$ is given by the formula

$$\mathcal{F}[\mathcal{M}[\varphi(x_0, x_1, x_2, x_3)]] = C \varphi(-x_0, x_1, x_2, x_3),$$

which is equally valid for integrable functions as well as distributions.

(ii) Note that $C = 1$, i.e. $d^4x = d^4\mu(x)$. Indeed, take $\varphi(x)$ to be the characteristic function of $\mathbb{Z}_p^4$, now

$$\varphi(x) = \mathcal{M}\left[\mathcal{M}\left[\varphi\right]\right](x) = C\mathcal{M}_{k \to x}\left[\mathcal{F}\left[\varphi\right](k_0, -\boldsymbol{k})\right] = C\mathcal{M}_{k \to x}\left[\varphi(k_0, \boldsymbol{k})\right]$$

$$= C^2 \mathcal{F}\left[\varphi(x_0, -\boldsymbol{x})\right] = C^2\varphi(x),$$

therefore $C = 1$.

6.2.3 Invariant Measures Under the Orthogonal Group $O(Q)$

We set $Q(x) = [x, x]$ as before. We also set

$$G := \begin{bmatrix} 1 & 0 & 0 & 0 \\ 0 & -1 & 0 & 0 \\ 0 & 0 & -1 & 0 \\ 0 & 0 & 0 & -1 \end{bmatrix}.$$

Then $Q(x) = x^T G x$, where T denotes the transpose of a matrix. The orthogonal group of $Q(x)$ is defined as

$$O(Q) = \left\{ \Lambda \in GL_4\left(\mathbb{Q}_p\right) ; [\Lambda x, \Lambda y] = [x, y] \right\}$$

$$= \left\{ \Lambda \in GL_4\left(\mathbb{Q}_p\right) ; \Lambda^T G \Lambda = G \right\}.$$

Notice that any $\Lambda \in O(Q)$ satisfies $\det \Lambda = \pm 1$. We consider $O(Q)$ as a p-adic Lie subgroup of $GL_4\left(\mathbb{Q}_p\right)$, which is also a p-adic Lie group.

For $t \in \mathbb{Q}_p^{\times}$, we set

$$V_t := \left\{ k \in \mathbb{Q}_p^4 ; Q(k) = t \right\}.$$

Proposition 153 (Rallis-Schiffman [92, Proposition 2–2]) *The orthogonal group $O(Q)$ acts transitively on V_t. On each orbit V_t there is a measure which is invariant under $O(Q)$ and unique up to multiplication by a positive constant.*

For each $t \in \mathbb{Q}_p^{\times}$, let $d\mu_t$ be a measure on V_t invariant under $O(Q)$. Since V_t is closed in $\mathbb{Q}_p^4$, it is possible to consider $d\mu_t$ as a measure on $\mathbb{Q}_p^4$ supported on V_t.

6.2.4   Some Remarks About p-Adic Analytic Manifolds

We need some basic results about p-adic manifolds in the sense of Serre, the reader may consult Chap. 2 or [57, 104].

Since $\nabla Q(k) \neq 0$ for any $k \in V_t$, by using the non-Archimedean implicit function theorem, one verifies that V_t is *a p-adic closed submanifold of codimension 1*, i.e a p-adic analytic hypersurface without singularities. The condition $\nabla Q(k) \neq 0$ for any $k \in V_t$, implies the existence of Gel'fand-Leray differential form λ_t on V_t, i.e.

$$dk_0 \wedge dk_1 \wedge dk_2 \wedge dk_3 = dQ(k) \wedge \lambda_t. \tag{6.4}$$

We denote the corresponding measure as $|\lambda_t|(A)$ for an open compact subset A of V_t.

The notation $\left(k_0, \ldots, \hat{k}_{l(j)}, \ldots, k_3\right)$ means omit the $l(j)$th coordinate. We now describe this measure in a suitable chart. We may assume that V_t is a countable disjoint union of submanifolds of the form

$$V_t^{(j)} := \left\{ \begin{array}{l} (k_0, \ldots, k_3) \in \mathbb{Q}_p^4; k_{l(j)} = h_j\left(k_0, \ldots, \hat{k}_{l(j)}, \ldots, k_3\right) \\ \text{with } \left(k_0, \ldots, \hat{k}_{l(j)}, \ldots, k_3\right) \in V_j, \end{array} \right\} \tag{6.5}$$

where $h_j\left(k_0, \ldots, \hat{k}_{l(j)}, \ldots, k_3\right)$ is a p-adic analytic function on some open compact subset V_j of $\mathbb{Q}_p^3$, and $\frac{\partial Q}{\partial k_{l(j)}}(z) \neq 0$ for any $z \in V_t^{(j)}$. If A is a compact open subset contained in $V_t^{(j)}$, then

$$|\lambda_t|(A) = \int_{h_j^{-1}(A)} \frac{dk_0 \ldots d\hat{k}_{l(j)} \ldots dk_3}{\left| \frac{\partial Q}{\partial k_{l(j)}}(k) \right|_p}, \tag{6.6}$$

where we are identifying the set $A \subset V_t^{(j)}$ with the set of all the coordinates of the points of A, which is a subset of $\mathbb{Q}_p^3$, and $h_j^{-1}(A)$ denotes the subset of $\mathbb{Q}_p^4$ consisting of the points (k_0, k_1, k_2, k_3) such that $k_{l(j)} = h_j\left(k_0, \ldots, \hat{k}_{l(j)}, \ldots, k_3\right)$ for $\left(k_0, \ldots, \hat{k}_{l(j)}, \ldots, k_3\right) \in A$.

Remark 154

(i) Let $\mathcal{T}(V_t)$ denote the family of all compact open subsets of V_t. Then λ_t is a additive function on $\mathcal{T}(V_t)$ such that $\lambda_t(A) \geq 0$ for every A in $\mathcal{T}(V_t)$. By Carath éodory's extension theorem λ_t has a unique extension to the σ-algebra generated by $\mathcal{T}(V_t)$. We also note that the measure λ_t is supported on V_t.

(ii) Let $\mathcal{D}(V_t)$ denote the $\mathbb{C}$-vector space generated by the characteristic functions of the elements of $\mathcal{T}(V_t)$. The fact that λ_t is a positive additive function on

$\mathcal{T}(V_t)$ is equivalent to say that

$$\begin{aligned} \mathcal{D}(V_t) &\to \quad \mathbb{C} \\ \varphi &\to \int_{V_t} \varphi\, |\lambda_t| \end{aligned}$$

is a positive distribution. We can identify the measure $|\lambda_t|$ with a distribution on $\mathbb{Q}_p^4$ supported on V_t.

(iv) Some authors use $\delta\left(Q(k) - t\right)$ or $\delta\left(Q(k) - t\right) d^4k$ to denote the measure $|\lambda_t|$. We will use $\delta\left(Q(k) - t\right)$.

Remark 155 Let G_0 be a subgroup of $GL_4(\mathbb{Q}_p)$. Let $\varphi \in \mathcal{D}\left(\mathbb{Q}_p^4\right)$ and let $\Lambda \in G_0$. We define the action of Λ on φ by putting

$$(\Lambda\varphi)(x) = \varphi\left(\Lambda^{-1}x\right),$$

and the action of Λ on a distribution $T \in \mathcal{D}'\left(\mathbb{Q}_p^4\right)$ by putting

$$(\Lambda T, \varphi) = \left(T, \Lambda^{-1}\varphi\right).$$

We say that T is invariant under G_0 if $\Lambda T = T$ for any $\Lambda \in G_0$.

Lemma 156 *With the above notation, we have $d\mu_t = C\,|\lambda_t|$ for some positive constant C.*

Proof By Remark 154 and Proposition 153, it is sufficient to show that the distribution $\delta\left(Q(k) - t\right)$ is invariant under $O(Q)$, i.e.

$$\int_{V_t} \varphi\left(\Lambda k\right) |\lambda_t(k)| = \int_{V_t} \varphi(k)\, |\lambda_t(k)|$$

for any $\Lambda \in O(Q)$ and $\varphi \in \mathcal{D}\left(\mathbb{Q}_p^4\right)$. Now, since V_t is invariant under Λ, it is sufficient to show that $|\lambda_t(k)| = |\lambda_t(y)|$ under $k = \Lambda^{-1}y$, for any $\Lambda \in O(Q)$. To verify this fact we note that

$$dk_0 \wedge dk_1 \wedge dk_2 \wedge dk_3 = \left(\det \Lambda^{-1}\right) dy_0 \wedge dy_1 \wedge dy_2 \wedge dy_3 \text{ and } dQ(k) = dQ(y)$$

under $k = \Lambda^{-1}y$. Now by (6.4) and the fact that the restriction of λ_t to V_t is unique we have $\lambda_t(k) = (\det \Lambda)\, \lambda_t(y)$ on V_t, i.e. $|\lambda_t(k)| = |\lambda_t(y)|$ under $k = \Lambda^{-1}y$ on V_t. ∎

6.2.5 Some Additional Results on $\delta\left(Q\left(k\right)-t\right)$

We now take $t = m^2$ with $m \in \mathbb{Q}_p^{\times}$. Notice that V_{m^2} has infinitely many points and that $(k_0, \boldsymbol{k}) \in V_{m^2}$ if and only if $(-k_0, \boldsymbol{k}) \in V_{m^2}$. In order to exploit this symmetry we need a 'notion of positivity' on $\mathbb{Q}_p$. To motivate our definitions consider $a = p^{-n}a_{-n} + p^{-n+1}a_{-n+1} + \cdots = p^{-n}ac\,(a) \in \mathbb{Q}_p^{\times}$, with $ac\,(a) \in \mathbb{Z}_p^{\times}$, i.e. $a_{-n} \neq 0$, then

$$-a = (p - a_{-n})\, p^{-n} + (p - 1 - a_{-n+1})\, p^{-n+1} + \cdots + (p - 1 - a_0)$$
$$+ (p - 1 - a_1)\, p + \cdots = p^{-n}ac(-a).$$

Thus, changing the sign of a is equivalent to changing the sign of its angular component. On the other hand, the equation $x^2 = a$ has two solutions if and only if n is even and $\left(\frac{a_{-n}}{p}\right) = 1$, here $\left(\frac{\cdot}{p}\right)$ denotes the Legendre symbol. The condition $\left(\frac{a_{-n}}{p}\right) = 1$ means that the equation $z^2 \equiv a_{-n} \bmod p$ has two solutions, say $\pm z_0$, because $p \neq 2$, with $z_0 \in \left\{1, \ldots, \frac{p-1}{2}\right\}$ and $-z_0 \in \left\{\frac{p+1}{2}, \ldots, p - 1\right\}$.

We define

$$\mathbb{F}_p^{+} = \left\{1, \ldots, \frac{p-1}{2}\right\} \subset \mathbb{F}_p^{\times} \text{ and } \mathbb{F}_p^{-} = \left\{\frac{p+1}{2}, \ldots, p - 1\right\} \subset \mathbb{F}_p^{\times}.$$

Motivated by the above discussion we introduce the following notion of 'positivity'.

Definition 157 We say that $a \in \mathbb{Q}_p^{\times}$ is positive if $\overline{ac(a)} \in \mathbb{F}_p^{+}$, otherwise we declare a to be negative. We will use the notation $a > 0$, in the first case, and $a < 0$ in the second case.

The reader must be aware that this notion of positivity is not compatible with the arithmetic operations on $\mathbb{F}_p$ neither on $\mathbb{Q}_p^{\times}$ because these fields cannot be ordered.

We now define the *mass shells* as follows:

$$V_{m^2}^{+} = \{(k_0, \boldsymbol{k}) \in V_{m^2}; k_0 > 0\} \text{ and } V_{m^2}^{-} = \{(k_0, \boldsymbol{k}) \in V_{m^2}; k_0 < 0\}.$$

Hence

$$V_{m^2} = V_{m^2}^{+} \bigsqcup V_{m^2}^{-} \bigsqcup \{(k_0, \boldsymbol{k}) \in V_{m^2}; k_0 = 0\}. \tag{6.7}$$

Notice that

$$V_{m^2}^{+} \quad \to \quad V_{m^2}^{-}$$

$$(k_0, \boldsymbol{k}) \to (-k_0, \boldsymbol{k})$$

is a bijection. We define

$$\Pi : \quad \mathbb{Q}_p^4 \quad \to \mathbb{Q}_p^3$$

$$(k_0, \boldsymbol{k}) \to \boldsymbol{k},$$

and $\Pi\left(V_{m^2}^+\right) = \Pi\left(V_{m^2}^-\right) := U_{Q,m}$. Given $\boldsymbol{k} \in U_{Q,m}$, there are two p-adic numbers, $k_0 > 0$ and $-k_0 < 0$, such that $(k_0, \boldsymbol{k})$, $(-k_0, \boldsymbol{k}) \in V_{m^2}$, thus we can define the following two functions:

$$U_{Q,m} \to \qquad \mathbb{Q}_p^{\times} \qquad U_{Q,m} \to \qquad \mathbb{Q}_p^{\times}$$
$$\boldsymbol{k} \to \sqrt{\boldsymbol{k} \cdot \boldsymbol{k} + m^2} =: k_0 \qquad \boldsymbol{k} \to -\sqrt{\boldsymbol{k} \cdot \boldsymbol{k} + m^2} =: -k_0.$$

Furthermore, we obtain the following description of the sets $V_{m^2}^{\pm}$:

$$V_{m^2}^{\pm} = \left\{ (k_0, \boldsymbol{k}) \in \mathbb{Q}_p^4; k_0 = \pm\sqrt{\boldsymbol{k} \cdot \boldsymbol{k} + m^2}, \text{ for } \boldsymbol{k} \in U_{Q,m} \right\}. \tag{6.8}$$

Lemma 158 *With the above notation the following assertions hold:*

(i) $U_{Q,m}$ is an open subset of $\mathbb{Q}_p^3$;
(ii) the functions $\pm\sqrt{\boldsymbol{k} \cdot \boldsymbol{k} + m^2}$ are p-adic analytic on $U_{Q,m}$;
(iii) $U_{Q,m}$ is p-adic bianalytic equivalent to each $V_{m^2}^{\pm}$, and $V_{m^2}^{\pm}$ are open subsets of $\mathbb{Q}_p^3$;
(iv)

$$\lambda_{m^2} (k_0, \boldsymbol{k}) \mid_{V_{m^2}^{\pm}} = \frac{dk_1 \wedge dk_2 \wedge dk_3}{\pm 2\sqrt{\boldsymbol{k} \cdot \boldsymbol{k} + m^2}} \mid_{U_{Q,m}} \text{ and}$$

$$|\lambda_{m^2} (k_0, \boldsymbol{k})| \mid_{V_{m^2}^{\pm}} = \frac{d^3\boldsymbol{k}}{\left| \sqrt{\boldsymbol{k} \cdot \boldsymbol{k} + m^2} \right|_p} \mid_{U_{Q,m}},$$

where $d^3\boldsymbol{k}$ is the normalized Haar measure of $\mathbb{Q}_p^3$;

(v) $\displaystyle\int\limits_{\{(k_0, \boldsymbol{k}) \in V_{m^2}; k_0 = 0\}} \varphi (k_0, \boldsymbol{k}) |\lambda_{m^2} (k_0, \boldsymbol{k})| = 0$ *for any $\varphi (k_0, \boldsymbol{k}) \in \mathcal{D}\left(\mathbb{Q}_p^4\right)$.*

Proof Take a point $(k_0, \boldsymbol{k}) \in V_{m^2}^+$, then $(k_0, \boldsymbol{k}) \in V_t^{(j)}$ for some j, see (6.5), thus there exist an open compact subset $U_+ = U_+' \times U_+''$ containing $(k_0, \boldsymbol{k})$ and a p-adic analytic function $h_+ : U_+'' \to U_+'$ such that

$$V_{m^2}^+ \cap U = \{(k_0, \boldsymbol{k}) \in U_+; k_0 = h_+ (\boldsymbol{k}) \text{ with } \boldsymbol{k} \in U_+''\}. \tag{6.9}$$

Now by (6.8), we have

$$h_+ (k) \mid_{U_+''} = \sqrt{k \cdot k + m^2} \mid_{U_+''},$$

which implies that $\sqrt{k \cdot k + m^2}$ is a p-adic analytic function on $U_{Q,m}$, which is an open subset of $\mathbb{Q}_p^3$ since it is the union of all the U_+'' which are open. In this way we establish (i)–(ii).

We now prove (iii). By (ii)

$$U_{Q,m} \rightarrow \qquad\qquad V_{m^2}^{\pm}$$

$$k \rightarrow \left(\pm \sqrt{k \cdot k + m^2}, k \right) =: i_{\pm} (k)$$

are p-adic bianalytic mappings, and by (i) $V_{m^2}^{\pm}$ are open subsets of $\mathbb{Q}_p^3$.

The formulas (iv)–(v) follow from (6.4) by a direct calculation. ∎

Remark 159 Let X be a locally compact and totally disconnected topological space. We denote by $\mathcal{D}(X)$ the set of all complex-valued and locally constant functions on X. Any such function is a linear combination of characteristic functions of compact open sets. The strong dual space $\mathcal{D}'(X)$ of $\mathcal{D}(X)$ agrees with the algebraic dual of $\mathcal{D}(X)$. For further details the reader may consult [57, Chapter 7].

Lemma 160

(i) *Each of the spaces $\mathcal{D}(V_{m^2}^{\pm})$ is isomorphic to $\mathcal{D}(U_{Q,m})$ as $\mathbb{C}$-vector space.*
(ii) *If $\phi : U_{Q,m} \rightarrow \mathbb{C}$ is a function with compact support, then*

$$\int_{U_{Q,m}} \phi (k) \, \frac{d^3 k}{\left| \sqrt{k \cdot k + m^2} \right|_p} = \int_{V_{m^2}^{\pm}} \left(\phi \circ i_{\pm}^{-1} \right) (k) \, |\lambda_{m^2} (k_0, k)| .$$

(iii) *If $m \in \mathbb{Q}_p^{\times}$ and $\varphi : \mathbb{Q}_p^4 \rightarrow \mathbb{C}$ is a function with compact support, then*

$$\int_{V_{m^2}} \varphi (k) \, |\lambda_{m^2} (k)| = \int_{U_{Q,m}} \varphi \left(\sqrt{k \cdot k + m^2}, k \right) \frac{d^3 k}{\left| \sqrt{k \cdot k + m^2} \right|_p}$$

$$+ \int_{U_{Q,m}} \varphi \left(-\sqrt{k \cdot k + m^2}, k \right) \frac{d^3 k}{\left| \sqrt{k \cdot k + m^2} \right|_p}. \qquad (6.10)$$

Proof (i) Let $\phi_{\pm}$ be a function in $\mathcal{D}(V_{m^2}^{\pm})$, by applying Lemma 158 (iii), we have $\phi_{\pm} \circ i_{\pm} \in \mathcal{D}(U_{Q,m})$. Conversely, if $\varphi \in \mathcal{D}(U_{Q,m})$, then, by Lemma 158 (iii), $\varphi \circ i_{\pm}^{-1} \in \mathcal{D}(V_{m^2}^{\pm})$. (ii) The formula follows from (i) by applying Lemma 158 (iv). (iii) The formula follows from (6.7) by applying (ii) and Lemma 158 (v). ∎

Remark 161 We set

$$\left(\delta_\pm \left(Q\left(k\right) - m^2\right), \varphi\right) := \int_{V_{m^2}^\pm} \varphi\left(k\right) \left|\lambda_{m^2}\left(k\right)\right|, \ \varphi \in \mathcal{D}(\mathbb{Q}_p^4).$$

Then $\delta_\pm \left(Q\left(k\right) - m^2\right) \in \mathcal{D}'(\mathbb{Q}_p^4)$ and by(6.10),

$$\delta\left(Q\left(k\right) - m^2\right) = \delta_+\left(Q\left(k\right) - m^2\right) + \delta_-\left(Q\left(k\right) - m^2\right).$$

Now, if we take $\Lambda_0 := \begin{bmatrix} -1 & 0 \\ 0 & I_{3\times3} \end{bmatrix} = \Lambda_0^{-1} \in O\left(Q\right)$, then

$$\delta_-\left(Q\left(k\right) - m^2\right) = \Lambda_0 \delta_+\left(Q\left(k\right) - m^2\right).$$

Notice that instead of Λ_0 we can use any Λ satisfying $\Lambda\left(V_{m^2}^+\right) = V_{m^2}^-$.

6.2.6 The p-Adic Restricted Lorentz Group

Definition 162 We define the p-adic restricted Lorentz group $\mathcal{L}_+^\uparrow$ to be the largest subgroup of $SO(Q)$ such that $\mathcal{L}_+^\uparrow \left(V_{m^2}^\pm\right) = V_{m^2}^\pm$.

Notice that $\mathcal{L}_+^\uparrow$ is a non-trivial subgroup of $SO(Q)$. Indeed, take Λ in

$$SO\left(3\right) = \left\{R \in GL_3\left(\mathbb{Q}_p\right); R^T = R^{-1}, \ \det R = 1\right\},$$

and define

$$\tilde{\Lambda} = \begin{bmatrix} 1 & 0 \\ 0 & \Lambda \end{bmatrix}.$$

Then

$$\left(\tilde{\Lambda}\right)^T = \begin{bmatrix} 1 & 0 \\ 0 & \Lambda^{-1} \end{bmatrix} \text{ and } \left(\tilde{\Lambda}\right)^T G \tilde{\Lambda} = G, \text{ i.e. } \tilde{\Lambda} \in SO(Q),$$

and since $\tilde{\Lambda} \boldsymbol{k} \cdot \tilde{\Lambda} \boldsymbol{k} = \boldsymbol{k} \cdot \boldsymbol{k}$ we have $\tilde{\Lambda}\left(V_{m^2}^\pm\right) = V_{m^2}^\pm$.

At the moment, we do not know if $\mathcal{L}_+^\uparrow = \{1\} \times SO(3)$. It seems that this depends on $\mathbb{Q}_p$, which can be replaced for any locally compact field of characteristic different from 2.

Lemma 163 *The distributions* $\delta_\pm \left(Q(k) - m^2 \right)$ *are invariant under* $\mathcal{L}_+^\uparrow$.

Proof Take $\Lambda \in \mathcal{L}_+^\uparrow$, then

$$\left(\Lambda \delta_\pm \left(Q(k) - m^2 \right), \varphi \right) = \int_{V_{m^2}^\pm} \varphi(\Lambda k) \left| \lambda_{m^2}(k) \right| = \int_{V_{m^2}^\pm} \varphi(k) \left| \lambda_{m^2}(k) \right|$$

because $\Lambda \left(V_{m^2}^\pm \right) = V_{m^2}^\pm$ and $d\lambda_{m^2}$ is invariant under any element of $O(Q)$, see proof of Lemma 156. ∎

6.3 A *p*-Adic Analog of the Klein-Gordon Equation

Given a positive real number α and a nonzero *p*-adic number m, we define the pseudodifferential operator

$$\mathcal{D} \left(\mathbb{Q}_p^4 \right) \to C \left(\mathbb{Q}_p^4 \right) \cap L^2 \left(\mathbb{Q}_p^4 \right)$$

$$\varphi \quad \to \quad \square_{\alpha,m} \varphi,$$

where $(\square_{\alpha,m} \varphi)(x) := \mathcal{M}_{k \to x}^{-1} \left[\left| [k,k] - m^2 \right|_p^\alpha \mathcal{M}_{x \to k} \varphi \right]$.

We set $\mathcal{E}_{Q,m} \left(\mathbb{Q}_p^4 \right) := \mathcal{E}_{Q,m}$ to be the subspace of $\mathcal{D}' \left(\mathbb{Q}_p^4 \right)$ consisting of the distributions T such that the product $\left| [k,k] - m^2 \right|_p^\alpha \mathcal{M} T$ exists in $\mathcal{D}' \left(\mathbb{Q}_p^4 \right)$, here $\left| [k,k] - m^2 \right|_p^\alpha$ denotes the distribution $\varphi \to \int_{\mathbb{Q}_p^4} \left| [k,k] - m^2 \right|_p^\alpha \varphi(k) \, d^4 k$. Notice that $\mathcal{E} \left(\mathbb{Q}_p^4 \right)$, the space of locally constant functions, is contained in $\mathcal{E}_{Q,m}$. We consider $\mathcal{E}_{Q,m}$ as topological space with the topology inherited from $\mathcal{D}' \left(\mathbb{Q}_p^4 \right)$.

Definition 164 *A weak solution of*

$$\square_{\alpha,m} T = S, \text{ with } S \in \mathcal{D}' \left(\mathbb{Q}_p^4 \right), \tag{6.11}$$

is a distribution $T \in \mathcal{E}_{Q,m} \left(\mathbb{Q}_p^4 \right)$ *satisfying* (6.11).

For a subset U of $\mathbb{Q}_p^4$ we denote by 1_U its characteristic function.

Lemma 165 *Let* $T, S \in \mathcal{D}'\left(\mathbb{Q}_p^4\right)$. *The following assertions are equivalent:*

(i) *there exists* $W \in \mathcal{D}'\left(\mathbb{Q}_p^4\right)$ *such that* $TS = W$;
(ii) *for each* $x \in \mathbb{Q}_p^4$, *there exists an open compact subset* U *containing* x *so that for each each* $k \in \mathbb{Q}_p^4$:

$$\mathcal{M}\left[1_U W\right](k) := \int_{\mathbb{Q}_p^4} \mathcal{M}\left[1_U T\right](l)\, \mathcal{M}\left[1_U S\right](k - l)\, d^4 l$$

exists.

Proof Any distribution is uniquely determined by its restrictions to any countable open covering of $\mathbb{Q}_p^n$, see e.g. [111, p. 89]. On the other hand, the product TS exists if and only if $\mathcal{M}[T] * \mathcal{M}[S]$ exists, and in this case $\mathcal{M}[TS] = \mathcal{M}[T] * \mathcal{M}[S]$, see e.g. [111, p. 115]. Assume that $TS = W$ exists and take a countable covering $\{U_i\}_{i \in \mathbb{N}}$ of $\mathbb{Q}_p^4$ by open and compact subsets, then $TS\,|_{U_i} = W\,|_{U_i}$ i.e. $1_{U_i} TS = 1_{U_i} W$. We recall that the product of a finite number of distributions involving at least one distribution with compact support is associative and commutative, see e.g. [105, Theorem 3.19], then

$$W\,|_{U_i} = 1_{U_i} TS = 1_{U_i}\left(1_{U_i} TS\right) = \left(1_{U_i} T\right)\left(1_{U_i} S\right) = T\,|_{U_i}\, S\,|_{U_i}.$$

Now for each $x \in \mathbb{Q}_p^4$, there exists an open compact subset U_i containing x such that $\mathcal{M}\left[T\,|_{U_i}\right] * \mathcal{M}\left[S\,|_{U_i}\right] = \mathcal{M}\left[1_{U_i} T\right] * \mathcal{M}\left[1_{U_i} S\right] = \mathcal{M}\left[1_{U_i} W\right]$. Conversely, if for each x there exists an open compact subset U_i containing x (from this we get countable subcovering of $\mathbb{Q}_p^4$ also denoted as $\{U_i\}_{i \in \mathbb{N}}$) such that $\mathcal{M}\left[T\,|_{U_i}\right] * \mathcal{M}\left[S\,|_{U_i}\right] = \mathcal{M}\left[W\,|_{U_i}\right]$ i.e. $T\,|_{U_i}\, S\,|_{U_i} = W\,|_{U_i}$ exists, then $TS = W$. ∎

Corollary 166 *If* TS *exists, then* $supp(TS) \subseteq supp\,(T) \cap supp(S)$.

Proof Since $x \notin supp(S)$, there exists a compact open set U containing x such $(S, \varphi) = 0$ for any $\varphi \in \mathcal{D}(U)$, hence $1_U S = 0$, and $\mathcal{M}\left[1_U T\right] * \mathcal{M}\left[1_U S\right] = 0 = \mathcal{M}\left[1_U W\right]$, i.e. $W\,|_U = 0$, which means $x \notin supp(W)$. ∎

Remark 167 Lemma 165 and Corollary 166 are valid in arbitrary dimension. These results are well-known in the Archimedean setting, see e.g. [96, Theorem IX.43], however, such results do not appear in the standard books of p-adic analysis, see e.g. [5, 80, 105, 111].

Remark 168

(i) Let Ω denote the characteristic function of the interval $[0, 1]$. Then $\Omega\left(p^{-j}\|x\|_p\right)$ is the characteristic function of the ball $B_j^{(n)}(0)$. Let us recall definition of the product of two distributions. Set $\delta_j(x) := p^{nj}\Omega\left(p^j\|x\|_p\right)$ for $j \in \mathbb{N}$. Given

$T, S \in \mathcal{D}'\left(\mathbb{Q}_p^n\right)$, their product TS is defined by

$$(TS, \varphi) = \lim_{j \to +\infty} \left(S, \left(T * \delta_j\right)\varphi\right)$$

if the limit exists for all $\varphi \in \mathcal{D}\left(\mathbb{Q}_p^n\right)$.

(ii) We assert that

$$\left(\left|[k, k] - m^2\right|_p^{\alpha} \mathcal{M}T, \varphi\right) = \left(\mathcal{M}T, \left|[k, k] - m^2\right|_p^{\alpha} \varphi\right)$$

for any $T \in \mathcal{E}_{Q,m}\left(\mathbb{Q}_p^4\right)$ and any $\varphi \in \mathcal{D}\left(\mathbb{Q}_p^n\right)$. Indeed, by using the fact that V_{m^2} has $d^4 y$-measure zero,

$$\left|[k, k] - m^2\right|_p^{\alpha} * \delta_j(k) = \left(\left|[y, y] - m^2\right|_p^{\alpha}, \delta_j(k - y)\right)$$

$$= p^{4j} \int_{k + (p^j \mathbb{Z}_p)^4} \left|[y, y] - m^2\right|_p^{\alpha} d^4 y$$

$$= p^{4j} \int_{k + (p^j \mathbb{Z}_p)^4 \setminus V_{m^2}} \left|[y, y] - m^2\right|_p^{\alpha} d^4 y = \left|[k, k] - m^2\right|_p^{\alpha}$$

for j big enough depending on k. Then

$$\left(\left|[k, k] - m^2\right|_p^{\alpha} \mathcal{M}T, \varphi\right) = \lim_{j \to +\infty} \left(\mathcal{M}T(k), \left[\left|[k, k] - m^2\right|_p^{\alpha} * \delta_j(k)\right]\varphi(k)\right)$$

$$= \left(\mathcal{M}T(k), \left|[k, k] - m^2\right|_p^{\alpha} \varphi(k)\right).$$

Lemma 169 *A distribution $T \in \mathcal{E}_{Q,m}\left(\mathbb{Q}_p^4\right)$ is a weak solution of $\square_{\alpha,m} T = 0$ if and only if supp$\mathcal{M}T \subseteq V_{m^2}$.*

Proof Suppose that supp$\mathcal{M}T \subseteq V_{m^2}$, then by Corollary 166, we have

$$\text{supp}\left(\left|[k, k] - m^2\right|_p^{\alpha} \mathcal{M}T\right) \subseteq \text{supp}(\mathcal{M}T) \cap \text{supp}\left(\left|[k, k] - m^2\right|_p^{\alpha}\right) = \emptyset$$

because $\left|[k, k] - m^2\right|_p^{\alpha} = \left|[k, k] - m^2\right|_p^{\alpha} 1_{\mathbb{Q}_p^4 \setminus V_{m^2}}$ in $\mathcal{D}'(\mathbb{Q}_p^4)$ (V_{m^2} has $d^4 k$-measure zero) and supp$\left(\left|[k, k] - m^2\right|_p^{\alpha} 1_{\mathbb{Q}_p^4 \setminus V_{m^2}}\right) \subseteq \mathbb{Q}_p^4 \setminus V_{m^2}$, therefore $\left|[k, k] - m^2\right|_p^{\alpha} \mathcal{M}T = 0$.

Suppose now that $\left| [k,k] - m^2 \right|_p^\alpha \mathcal{M} T = 0$. By contradiction, assume that $\operatorname{supp} \mathcal{M} T \not\subseteq V_{m^2}$. Then , there exists $s_0 \in \mathbb{Q}_p^4 \smallsetminus V_{m^2}$ and an open compact subset $U \subset \mathbb{Q}_p^4 \smallsetminus V_{m^2}$ containing s_0 such that $(\mathcal{M} T, 1_U) \neq 0$. By using Remark 168-(ii) and by shrinking U if necessary,

$$\left(\left| [k,k] - m^2 \right|_p^\alpha \mathcal{M} T, 1_U \right) = \left(\mathcal{M} T, \left| [k,k] - m^2 \right|_p^\alpha 1_U \right)$$

$$= \left| [s_0, s_0] - m^2 \right|_p^\alpha (\mathcal{M} T, 1_U) \neq 0,$$

contradicting $\left| [k,k] - m^2 \right|_p^\alpha \mathcal{M} T = 0$. ∎

Remark 170 Let $\varphi \in \mathcal{D} \left(\mathbb{Q}_p^4 \right)$ and let $\Lambda \in \mathcal{L}_+^\uparrow$, a Lorentz transformation. By using that $[k,x] = \frac{1}{2} \{ Q(k+x) - Q(k) - Q(x) \}$, we have

$$\mathcal{M} [\varphi (\Lambda x)] (k) = \mathcal{M} [\varphi] (\Lambda k)$$

and

$$\Lambda \mathcal{M} [T] = \mathcal{M} [\Lambda T].$$

Hence the Fourier transform preserves Lorentz invariance, or more generally, the Fourier transform preserves invariance under $O(Q)$.

Proposition 171 *The distributions*

$$\mathcal{M} [T] (k) = a \delta_+ \left(Q(k) - m^2 \right) + b \delta_- \left(Q(k) - m^2 \right), \ a, b \in \mathbb{C},$$

are weak solutions of $\square_{\alpha, m} T = 0$ invariant under $\mathcal{L}_+^\uparrow$.

Proof By Remark 170, it is sufficient to show that $\delta_\pm \left(Q(k) - m^2 \right)$ are invariant solutions of $\left| [k,k] - m^2 \right|_p^\alpha \mathcal{M} [T] = 0$, which follows from Lemmas 163–169. ∎

At this point we should mention that a similar results to Lemmas 163–169 and Proposition 171 are valid for the Archimedean Klein-Gordon equation, see e.g. [31] or [32, Chapter IV].

6.4 The Cauchy Problem for the Non-Archimedean Klein-Gordon Equation

In this section we study the Cauchy problem for the p-adic Klein-Gordon equations.

6.4.1   Twisted Vladimirov Pseudodifferential Operators

Let $\mathbb{C}_1^{\times}$ denote the multiplicative group of complex numbers having modulus one. Let $\pi_1 : \mathbb{Z}_p^{\times} \to \mathbb{C}_1^{\times}$ be a *non-trivial multiplicative character* of $\mathbb{Z}_p^{\times}$ with positive *conductor* k, i.e. k is the smallest positive integer such that $\pi_1 \mid_{1+p^k\mathbb{Z}_p} = 1$. Some authors call a such character a *unitary character* of $\mathbb{Z}_p^{\times}$. We extend π_1 to $\mathbb{Q}_p^{\times}$ by putting $\pi_1(x) := \pi_1(ac(x))$. A *quasicharacter* of $\mathbb{Q}_p^{\times}$ (some authors use *multiplicative character*) is a continuous homomorphism from $\mathbb{Q}_p^{\times}$ into $\mathbb{C}^{\times}$. Every quasicharacter has the form $\pi_s(x) = \pi_1(x)\,|x|_p^{s-1}$ for some complex number s.

The distribution associated with $\pi_s(x)$ has a meromorphic continuation to the whole complex plane given by

$$
(\pi_s(x), \varphi(x)) = \int\limits_{\mathbb{Z}_p} \pi_1(x)\,|x|_p^{s-1}\,\{\varphi(x) - \varphi(0)\}\,dx
$$

$$
+ \int\limits_{\mathbb{Q}_p \smallsetminus \mathbb{Z}_p} \pi_1(x)\,|x|_p^{s-1}\,\varphi(x)\,dx, \tag{6.12}
$$

see e.g. [111, p. 117].

On the other hand,

$$
\mathcal{F}[\pi_s](\xi) = \Gamma_p(s, \pi_1)\,\pi_1^{-1}(\xi)\,|\xi|_p^{-s}, \text{ for any } s \in \mathbb{C}, \tag{6.13}
$$

where

$$
\Gamma_p(s, \pi_1) = p^{sk} a_{p,k}(\pi_1),
$$

$$
a_{p,k}(\pi_1) = \int\limits_{\mathbb{Z}_p^{\times}} \pi_1(t)\,\chi_p(p^{-k}t)\,dt \text{ and } \left|a_{p,k}(\pi_1)\right| = p^{-\frac{k}{2}},
$$

see e.g. [111, p. 124].

Another useful formula is the following:

$$
(\pi_s(x), \varphi(x)) = \int\limits_{\mathbb{Q}_p} \frac{\pi_1(x)\,\{\varphi(x) - \varphi(0)\}}{|x|_p^{s+1}}\,dx, \text{ for } \mathrm{Re}(s) > 0, \text{ and } \varphi \in \mathcal{D}(\mathbb{Q}_p).
$$

$$
\tag{6.14}
$$

The formula follows from (6.12) by using that

$$
\int\limits_{\mathbb{Q}_p \smallsetminus \mathbb{Z}_p} \frac{\pi_1(x)}{|x|_p^{s+1}}\,dx = \left(\sum_{j=1}^{\infty} p^{-js}\right) \int\limits_{\mathbb{Z}_p^{\times}} \pi_1(y)\,dy = 0. \tag{6.15}
$$

Given $\alpha > 0$, we define *the twisted Vladimirov operator* by

$$\left(\tilde{\boldsymbol{D}}_x^\alpha \varphi\right)(x) = \mathcal{F}_{k\to x}^{-1}\left(\pi_1^{-1}(k)\,|k|_p^\alpha\,\mathcal{F}_{x\to k}(\varphi)\right) \text{ for } \varphi \in \mathcal{D}\left(\mathbb{Q}_p\right).$$

Notice that

$$\mathcal{D}\left(\mathbb{Q}_p\right) \to C\left(\mathbb{Q}_p, \mathbb{C}\right) \cap L^2$$

$$\varphi \quad \to \quad \left(\tilde{\boldsymbol{D}}_x^\alpha \varphi\right)$$

is a well-defined linear operator.

Lemma 172 *For $\alpha > 0$ and $\varphi \in \mathcal{D}\left(\mathbb{Q}_p\right)$, the following formula holds:*

$$\left(\tilde{\boldsymbol{D}}_x^\alpha \varphi\right)(x) = \frac{1}{\Gamma_p(-\alpha, \pi_1)} \int_{\mathbb{Q}_p} \frac{\pi_1(y)\{\varphi(x-y)-\varphi(x)\}}{|y|_p^{\alpha+1}}\,dy. \qquad (6.16)$$

Proof By using (6.13), we have

$$\left(\tilde{\boldsymbol{D}}_x^\alpha \varphi\right)(x) = \mathcal{F}_{\xi\to x}^{-1}\left(\pi_1^{-1}(k)\,|k|_p^\alpha\right)(x) * \varphi(x) = \frac{1}{\Gamma_p(-\alpha,\pi_1)}\pi_{-\alpha}(x)*\varphi(x)$$

$$= \frac{1}{\Gamma_p(-\alpha,\pi_1)}\,(\pi_{-\alpha}(y), \varphi(x-y))$$

$$= \frac{1}{\Gamma_p(-\alpha,\pi_1)}\int_{\mathbb{Z}_p}\frac{\pi_1(y)\{\varphi(x-y)-\varphi(x)\}}{|y|_p^{\alpha+1}}\,dy$$

$$+ \frac{1}{\Gamma_p(-\alpha,\pi_1)}\int_{\mathbb{Q}_p\setminus\mathbb{Z}_p}\frac{\pi_1(y)\,\varphi(x-y)}{|y|_p^{\alpha+1}}\,dx$$

$$= \frac{1}{\Gamma_p(-\alpha,\pi_1)}\int_{\mathbb{Z}_p}\frac{\pi_1(y)\{\varphi(x-y)-\varphi(x)\}}{|y|_p^{\alpha+1}}\,dy$$

$$+ \frac{1}{\Gamma_p(-\alpha,\pi_1)}\int_{\mathbb{Q}_p\setminus\mathbb{Z}_p}\frac{\pi_1(y)\{\varphi(x-y)-\varphi(x)\}}{|y|_p^{\alpha+1}}\,dy,$$

where we used (6.15).   ∎

Notice that the right-hand side of (6.16) makes sense for a wider class of functions. For instance, for any locally constant function $u(x)$ satisfying

$$\int_{\mathbb{Q}_p \smallsetminus \mathbb{Z}_p} \frac{|u(x)|}{|x|_p^{\alpha+1}} dx < \infty.$$

Another useful formula is the following:

Lemma 173 *For $\alpha > 0$, we have*

$$\pi_1^{-1}(x) |x|_p^{\alpha} = \frac{1}{\Gamma_p(-\alpha, \pi_1)} \int_{\mathbb{Q}_p} \frac{\pi_1(y)\{\chi_p(yx) - 1\}}{|y|_p^{\alpha+1}} dy \text{ in } \mathcal{D}'(\mathbb{Q}_p).$$

Proof The formula follows from (6.13) and (6.14). The proof is a simple variation of the proof given for the case in which π_1 is the trivial character, see e.g. [80, Proposition 2.3].   ∎

From now on we put

$$\pi_1(x) := i\left(\overline{\frac{ac(x)}{p}}\right) \text{ for } x \in \mathbb{Z}_p^{\times},$$

where $\bar{z}$ denotes the reduction mod p of $z \in \mathbb{Z}_p$, $\left(\frac{\cdot}{p}\right)$ denotes the Legendre symbol. Notice that the conductor of π_1 is 1 and $\pi_1(x) \in \{\pm i\}$, furthermore,

$$\left(\frac{-1}{p}\right) = \begin{cases} 1, & \text{if } p - 1 \text{ is divisible by 4} \\ -1 & \text{if } p - 3 \text{ is divisible by 4}. \end{cases}$$

6.4.2   The Cauchy Problem for the p-Adic Klein-Gordon Equation

In this section we take $x_0 = t$ and $(x_0, x) = (t, x) \in \mathbb{Q}_p \times \mathbb{Q}_p^3$. Our goal is to study the following Cauchy problem:

$$\begin{cases} (\Box_{\alpha,m} u)(t, x) = 0, & \text{(A)} \\[2mm] u(t, x)|_{t=0} = \psi_0(x), & \psi_0 \in \mathcal{D}(\mathbb{Q}_p^3) \text{ and } \mathcal{F}^{-1}[\psi_0] \in \mathcal{D}(U_{Q,m}) \text{ (B)} \\[2mm] \tilde{D}_x^{\alpha} u(t, x)|_{t=0} = \psi_1(x), & \psi_1 \in \mathcal{D}(\mathbb{Q}_p^3) \text{ and } \mathcal{F}^{-1}[\psi_1] \in \mathcal{D}(U_{Q,m}) \text{ (C)}. \end{cases}$$
$$(6.17)$$

Theorem 174 *Assume that $p-3$ is divisible by 4. Then Cauchy problem (6.17) has a weak solution given by*

$$u(t,x) = \int_{U_{Q,m}} \chi_p\left(-t\sqrt{k\cdot k + m^2} + x\cdot k\right) u_+(k)\, \frac{d^3 k}{\left|\sqrt{k\cdot k + m^2}\right|_p}$$

$$+ \int_{U_{Q,m}} \chi_p\left(t\sqrt{k\cdot k + m^2} + x\cdot k\right) u_-(k)\, \frac{d^3 k}{\left|\sqrt{k\cdot k + m^2}\right|_p}, \qquad (6.18)$$

where

$$u_+(k) = \frac{1}{2}\left\{\left|\sqrt{k\cdot k + m^2}\right|_p \mathcal{F}^{-1}[\psi_0](k) - \frac{i\pi_1\left(\sqrt{k\cdot k + m^2}\right)}{\left|\sqrt{k\cdot k + m^2}\right|_p^{\alpha-1}} \mathcal{F}^{-1}[\psi_1](k)\right\},$$

and

$$u_-(k) = \frac{1}{2}\left\{\left|\sqrt{k\cdot k + m^2}\right|_p \mathcal{F}^{-1}[\psi_0](k) + \frac{i\pi_1\left(\sqrt{k\cdot k + m^2}\right)}{\left|\sqrt{k\cdot k + m^2}\right|_p^{\alpha-1}} \mathcal{F}^{-1}[\psi_1](k)\right\}.$$

Proof We first show that $u(t,x)$ is a weak solution of (6.17)-(A). By applying Lemmas 158–160, we have

$$\int_{U_{Q,m}} \chi_p\left(\mp t\sqrt{k\cdot k + m^2} + x\cdot k\right) u_\pm(k)\, \frac{d^3 k}{\left|\sqrt{k\cdot k + m^2}\right|_p}$$

$$= \int_{V_{m^2}^\pm} \chi_p\left([(k_0,k),(-t,-x)]\right) u_\pm\left(i_\pm^{-1}(k)\right) d\lambda_{m^2}(k_0,k)$$

$$= \mathcal{M}_{(k_0,k)\to(t,x)}^{-1}\left[\left(u_\pm \circ i_\pm^{-1}\right)(k)\,\delta_\pm\left(Q(k)-m^2\right)\right],$$

whence

$$\mathcal{M}_{(t,x)\to(k_0,k)}[u(t,x)] = \left[\left(u_+ \circ i_+^{-1}\right)(k)\,\delta_+\left(Q(k)-m^2\right)\right]$$

$$+ \left[\left(u_- \circ i_-^{-1}\right)(k)\,\delta_-\left(Q(k)-m^2\right)\right].$$

By Lemma 169, $u(t, x)$ is a weak solution of (6.22), if

$$\text{supp}\left(u_\pm \circ i_\pm^{-1}\right)(k)\,\delta_\pm\left(Q(k) - m^2\right) \subseteq V_{m^2}.$$

This last condition is verified by applying Corollary 166 and the fact that

$$\text{supp}\left(u_\pm \circ i_\pm^{-1}\right) \subseteq V_{m^2}^\pm \subset V_{m^2} \text{ and } \text{supp}\left(\delta_\pm\left(Q(k) - m^2\right)\right) \subseteq V_{m^2}^\pm \subset V_{m^2}.$$

The verification of (6.17)-(B) is straight forward. To verify (6.17)-(C) we proceed as follows. By using Lemma 172, Fubini's theorem and Lemma 173, we get the following formula:

$$\tilde{D}_t^\alpha \left[\int_{U_{Q,m}} \chi_p\left(\mp t\sqrt{k \cdot k + m^2} + x \cdot k\right) u_\pm(k) \, \frac{d^3 k}{\left|\sqrt{k \cdot k + m^2}\right|_p} \right]$$

$$= \int_{U_{Q,m}} \frac{\chi_p\left(\mp t\sqrt{k \cdot k + m^2} + x \cdot k\right) u_\pm(k)}{\left|\sqrt{k \cdot k + m^2}\right|_p}$$

$$\times \left\{ \frac{1}{\Gamma_p(-\alpha, \pi_1)} \int_{\mathbb{Q}_p} \frac{\pi_1(y)\left\{\chi_p\left(\pm y\sqrt{k \cdot k + m^2}\right) - 1\right\}}{|y|_p^{\alpha+1}} dy \right\} d^3 k$$

$$= \pi_1^{-1}(\pm 1) \int_{U_{Q,m}} \chi_p\left(\mp t\sqrt{k \cdot k + m^2} + x \cdot k\right) \pi_1^{-1}\left(\sqrt{k \cdot k + m^2}\right)$$

$$\times \left|\sqrt{k \cdot k + m^2}\right|_p^{\alpha-1} u_\pm(k) \, d^3 k.$$

Now Condition (6.17)-(C) follows from the previous formula. ∎

Remark 175

(i) Note that the condition '$p - 3$ is divisible by 4' is required only to establish (6.17)-(C). At the moment, we do not know if this condition is necessary to have (6.17)-(C).

(ii) The parameter α does not have any influence on the solutions of (6.22).

Theorem 176 *The equation*

$$\left(\square_{\alpha,m} u\right)(t, x) = J(t, x), \quad J(t, x) \in \mathcal{D}\left(\mathbb{Q}_p^4\right) \tag{6.19}$$

admits the following weak solution:

$$u(t,x) = E_\alpha(t,x) * J(t,x)$$

$$+ \int_{U_{Q,m}} \chi_p\left(-t\sqrt{k \cdot k + m^2} + x \cdot k\right) u_+(k) \frac{d^3k}{\left|\sqrt{k \cdot k + m^2}\right|_p}$$

$$+ \int_{U_{Q,m}} \chi_p\left(t\sqrt{k \cdot k + m^2} + x \cdot k\right) u_-(k) \frac{d^3k}{\left|\sqrt{k \cdot k + m^2}\right|_p}, \qquad (6.20)$$

where $E_\alpha(t,x)$ is a distribution on $S(\mathbb{Q}_p^4)$ satisfying

$$\left|[k,k] - m^2\right|_p^\alpha \mathcal{M}_{\substack{t \to k_0 \\ x \to k}} [E_\alpha(t,x)] = 1 \text{ in } \mathcal{D}'(\mathbb{Q}_p^4), \qquad (6.21)$$

and $u_+(k)$, $u_-(k)$ are arbitrary functions in $\mathcal{D}(U_{Q,m})$.

Proof Like in the classical case a solution of (6.19) is computed as $u_0(t,x) + u_1(t,x)$ with $u_0(t,x)$ a particular solution of (6.19) and $u_1(t,x)$ a general solution of

$$(\square_{\alpha,m} u_1)(t,x) = 0. \qquad (6.22)$$

The existence of a fundamental solution for (6.19), i.e. a distribution $E_\alpha(t,x)$ is a distribution on $\mathcal{D}(\mathbb{Q}_p^4)$ such that $E_\alpha(t,x) * J(t,x)$ is a weak solution of (6.19), was established in Theorem 134 in Chap. 6. This fundamental solution satisfies (6.21). Finally, we verify that

$$u_1(t,x) = \int_{U_{Q,m}} \chi_p\left(-t\sqrt{k \cdot k + m^2} + x \cdot k\right) u_+(k) \frac{d^3k}{\left|\sqrt{k \cdot k + m^2}\right|_p}$$

$$+ \int_{U_{Q,m}} \chi_p\left(t\sqrt{k \cdot k + m^2} + x \cdot k\right) u_-(k) \frac{d^3k}{\left|\sqrt{k \cdot k + m^2}\right|_p},$$

satisfies (6.22) as in the proof of Theorem 174.   ∎

Like in the Archimedean case the non-Archimedean Klein-Gordon equations admit plane waves.

Lemma 177 *Let $(E,p) \in V_{m^2}^\pm$, i.e. $E = \pm\sqrt{p \cdot p + m^2}$. Then $u(t,x) = \chi_p([(t,x),(E,p)])$ is a weak solution of $(\square_{\alpha,m} u)(t,x) = 0$.*

Proof (i) Since $\mathcal{M}_{(t,x) \to (k_0,k)}[u(t,x)] = \delta(k_0 - E, k - p)$, the results follows from Lemma 169.   ∎

6.5 Further Results on the *p*-Adic Klein-Gordon Equation

In this section we change the notation slightly, this facilitates the comparison with the classical results and constructions. Set

$$\omega\,(\boldsymbol{k}) := \sqrt{\boldsymbol{k}\cdot\boldsymbol{k} + m^2}\ \text{for } \boldsymbol{k}\in U_{Q,m}.$$

The function $\omega\,(\boldsymbol{k})$ is a *p*-adic analytic function on $U_{Q,m}$, cf. Lemma 158, and $\omega\,(\boldsymbol{k}) \neq 0$ for any $\boldsymbol{k} \in U_{Q,m}$. Then, by Taylor formula, $|\omega\,(\boldsymbol{k})|_p$ is a locally constant function on $U_{Q,m}$, and if $\phi_{\pm} \in \mathcal{D}\left(U_{Q,m}\right)$, then $|\omega\,(\boldsymbol{k})|_p^{\pm1}\,\phi_{\pm}\,(\boldsymbol{k}) \in \mathcal{D}\left(U_{Q,m}\right)$. Then

$$\phi\,(t,\boldsymbol{x}) = \int_{U_{Q,m}} \chi_p\,(-\boldsymbol{x}\cdot\boldsymbol{k})\left\{\chi_p\,(-t\omega\,(\boldsymbol{k}))\,\phi_+\,(\boldsymbol{k}) + \chi_p\,(t\omega\,(\boldsymbol{k}))\,\phi_-\,(\boldsymbol{k})\right\}d^3\boldsymbol{k}$$

$$(6.23)$$

is a weak solution of $\left(\Box_{\alpha,m}\phi\right)(t,\boldsymbol{x}) = 0$ for any $\phi_{\pm} \in \mathcal{D}\left(U_{Q,m}\right)$. Note that if we replace $\chi_p\,(-\boldsymbol{x}\cdot\boldsymbol{k})$ by $\chi_p\,(\boldsymbol{x}\cdot\boldsymbol{k})$ in (6.23) we get another weak solution.

As in the classical case, for the quantum interpretation we specialize to 'positive energy solutions' which have $\phi_-\,(\boldsymbol{k}) = 0$. Then the solution is determined by a single complex valued initial condition. The solution (6.23) with initial condition

$$\Psi \in L^2_{Q,m} := L^2\left(\left\{\Phi \in L^2\left(\mathbb{Q}_p^3\right); \operatorname{supp}\,(\mathcal{F}\Phi) \subseteq U_{Q,m}\right\}, d^3\boldsymbol{k}\right)$$

-where the condition 'supp$(\mathcal{F}\Phi) \subseteq U_{Q,m}$' means that there exists a function Φ' in the equivalence class containing $\mathcal{F}\Phi$ such that supp$(\Phi') \subseteq U_{Q,m}$- is given by

$$\Psi\,(t,\boldsymbol{x}) = \int_{U_{Q,m}} \chi_p\,(-t\omega\,(\boldsymbol{k}) - \boldsymbol{x}\cdot\boldsymbol{k})\,(\mathcal{F}\Psi)\,(\boldsymbol{k})\,d^3\boldsymbol{k}. \qquad (6.24)$$

Set

$$U\,(t) : L^2_{Q,m} \to L^2\left(\mathbb{Q}_p^3\right)$$

$$\Psi \;\to\; \Psi\,(t,\boldsymbol{x}) = \mathcal{F}^{-1}_{\boldsymbol{k}\to\boldsymbol{x}}\left(\chi_p\,(-t\omega\,(\boldsymbol{k}))\,\mathcal{F}_{\boldsymbol{x}\to\boldsymbol{k}}\Psi\right),$$

for $t \in \mathbb{Q}_p$.

Lemma 178

(i) $U\,(t)$, $t \in \mathbb{Q}_p$ is a group of unitary operators on $L^2_{Q,m}$.
(ii) st. $-\lim_{t\to 0} U\,(t) = I$.

Proof It is a straightforward calculation.   ∎

Chapter 7
Final Remarks and Some Open Problems

The theory pseudodifferential equations over p-adic fields is just beginning. There are several open problems connecting these equations with number theory, probability and physics. Here we want to pose some of them.

7.1 Zeta Functions, Adelic Riesz Kernels and Heat Equations

We use all the notation introduced in Chap. 4. We propose to study the meromorphic continuation and the existence of functional equations, i.e. to compute explicitly the Fourier transform, of the following distributions:

$$\varsigma_{\mathbb{A}_f}(s, \varphi) := \int_{\mathbb{A}_f \setminus \{x \in \mathbb{A}_f : \|x\| = 0\}} \varphi(x) \|x\|^s \, dx_{\mathbb{A}_f}, \tag{7.1}$$

$\varphi \in \mathcal{D}(\mathbb{A}_f)$, $s \in \mathbb{C}$ with $\mathrm{Re}(s) > 0$, and

$$\varsigma_{\mathbb{A}}(s, \phi) := \int_{\mathbb{A} \setminus \{x \in \mathbb{A} : \|x_f\| + |x_\infty|_\infty = 0\}} \varphi(x_f, x_\infty) \{\|x_f\| + |x_\infty|_\infty\}^s \, dx_{\mathbb{A}}, \tag{7.2}$$

$\varphi \in \mathcal{D}(\mathbb{A})$, $s \in \mathbb{C}$ with $\mathrm{Re}(s) > 0$. Notice that $\|x\|^{\mathrm{Re}(s)}$, with $\mathrm{Re}(s) > 0$, is a continuous function on $\mathbb{A}_f$, because $(\mathbb{A}_f, \rho)$ is a complete metric space with $\rho(x, y) = \|x - y\|$, $x, y \in \mathbb{A}_f$, and $\int_{\mathbb{A}_f} |\varphi(x)| \, dx_{\mathbb{A}_f} < \infty$ due to the fact that $|\varphi(x)|$ is a linear combination of characteristic functions of balls and the $dx_{\mathbb{A}_f}$-measure of a ball is finite. Therefore $\varsigma_{\mathbb{A}_f}(s, \varphi)$ converges for $\mathrm{Re}(s) > 0$ and any $\varphi \in \mathcal{D}(\mathbb{A}_f)$. A similar reasoning works for $\varsigma_{\mathbb{A}}(s, \phi)$. These distributions are 'additive' adelic Riesz

© Springer International Publishing AG 2016
W.A. Zúñiga-Galindo, *Pseudodifferential Equations Over Non-Archimedean Spaces*, Lecture Notes in Mathematics 2174, DOI 10.1007/978-3-319-46738-2_7

kernels. Take $\varphi_0(x) = \prod_{p<\infty} 1_{\mathbb{Z}_p}(x_p)$, by analogy with the Tate thesis, $\varsigma_{\mathbb{A}_f}(s,\varphi_0)$ is an 'additive version' of the Riemann zeta function, more precisely,

$$\varsigma_{\mathbb{A}_f}(s,\varphi_0) = \sum_{j\geq 0}\sum_{p<\infty}\{\Phi(p^j) - \Phi(p^j_-)\}p^{js}, \ \mathrm{Re}(s) > 0,$$

where $\Phi(\cdot)$ denotes the second Chebyshev function. A natural problem consists in extending Tate's thesis to adelic Riesz kernels of types (7.1)–(7.2).

We extend $\|\cdot\|$ to $\mathbb{A}_f^n$, with $n \geq 1$, by taking $\|(x_1,\ldots,x_n)\| := \max_i \|x_i\|$, then $\mathbb{A}_f^n$ with the metric induced by $\|\cdot\|$ is a complete metric space. Fix a non-constant polynomial $f \in \mathbb{Z}[y_1,\ldots,y_n]$. For $(x_1,\ldots,x_n) \in \mathbb{A}_f^n$, with $x_i = \{x_{i,p}\}_{p<\infty}, x_{i,p} \in \mathbb{Q}_p$, we define

$$f(x_1,\ldots,x_n) = \{f(x_{1,p},\ldots,x_{n,p})\}_{p<\infty},$$

then f gives rise a function from $\mathbb{A}_f^n$ into $\mathbb{A}_f$. Denote by $d^n x_{\mathbb{A}_f}$ the product measure of n copies of $dx_{\mathbb{A}_f}$, and define $\mathcal{D}\left(\mathbb{A}_f^n\right)$ as the algebraic tensor product of n copies of $\mathcal{D}(\mathbb{A}_f)$. Notice that $\|f(x_1,\ldots,x_n)\|^s : \mathbb{A}_f^n \to \mathbb{C}$, with $\mathrm{Re}(s) > 0$, is a continuous function, the continuity can be verified by using sequences. We attach to (f,φ), $\varphi \in \mathcal{D}\left(\mathbb{A}_f^n\right)$ the following adelic Igusa type zeta function:

$$\varsigma_{\mathbb{A}_f^n}(s,f;\varphi) := \int_{\mathbb{A}_f^n \setminus f^{-1}(0)} \varphi(x)\|f(x_1,\ldots,x_n)\|^s \, d^n x_{\mathbb{A}_f} \text{ for } \mathrm{Re}(s) > 0.$$

A natural problem is to determine if $Z_{\mathbb{A}_f^n}(s,f;\varphi)$ admits a meromorphic continuation to the whole complex plane. It is interesting to mention that other adelic local zeta functions were introduced in [88] and [59].

Denote by $\mathbb{A}^\times$ the group of ideles of $\mathbb{A}$. We define the Haran-Vladimirov operator

$$\boldsymbol{D}_{\mathbb{A}\times}^\alpha \varphi = \underset{p\leq\infty}{\otimes} \frac{1}{c_p^{(\alpha)}} \boldsymbol{D}_p^\alpha \varphi_p,$$

where $\alpha > 0$, $c_p^{(\alpha)}$ is a suitable positive constant and $\varphi \in \mathcal{D}(\mathbb{A}^\times)$, for a precise definition the reader may consult [51]. In this article, Haran established a remarkable and deep connection between operators of type $\boldsymbol{D}_{\mathbb{A}\times}^\alpha$ and the zeros of the Riemann zeta function. We propose two problems connected with operators of type $\boldsymbol{D}_{\mathbb{A}\times}^\alpha$. First, consider the following parabolic-type equation on $\mathbb{A}^\times$:

$$\frac{\partial u(x,t)}{\partial t} + \boldsymbol{D}_{\mathbb{A}\times}^\alpha u(x,t) = 0, x \in \mathbb{A}^\times, t \geq 0.$$

Is this an ultradiffusion equation (i.e. is it possible to attach a Markov process to this equation)? The author believes that this problem is connected with the results of [20] and [72]. The connections between the Riemann zeta function and the heat equation are well-known and deep. A natural problem consists in studying the spectral zeta functions attached to non-Archimedean pseudodifferential operators. This line of research has been started in [27].

7.2   Fundamental Solutions and Operators of Bernstein-Sato Type

The existence of fundamental solutions for pseudodifferential operators with polynomial type symbols given in Chap. 5 is based on the existence of a meromorphic continuation for the Igusa local zeta function attached to symbol, this last fact is a consequence of Hironaka's resolution of singularities theorem [54]. A natural problem is to show the existence of fundamental solutions for pseudodifferential operators, with polynomial type symbols, without using resolution of singularities.

In the Archimedean setting the existence of a fundamental solution for a partial differential operator with constant coefficients was established, independently, by L. Ehrenpreis [40] and B. Malgrange [84], see also [15, 18, 55, 56, 84, 89, 99]. According to [89], the existing proofs of the Malgrange-Ehrenpreis theorem can be classified into three categories:

(A) proofs using the Hahn-Banach theorem;
(B) proofs by means of an explicit formulae;
(C) proofs by solution of a division problem.

The main difficulty in adapting the proofs of type A to the p-adic setting is the lack of a 'true p-adic analog' of the Schwartz space. The space $\mathcal{D}$ is not useful in context of p-adic pseudodifferential equations due to fact that it is not invariant under pseudo-differential operators. Recently, in [117] the author introduced a new function space that 'seems to be the true non-Archimedean analog' of Schwartz space. The proofs of type B bypass the problem created by the set of zeros of the symbol passing from $\mathbb{R}^n$ to $\mathbb{C}^n$. It is not clear to the author if this type of techniques can be adapted to the p-adic setting. The solution of the division problems can be obtained in different ways. By using Hironaka's resolution of singularities theorem [15]. The non-Archimdean version of this technique was presented in Chap. 5. By using a combination of algebraic and analytic techniques, including the notion of tempered distribution, as in [55]. Finally, by using algebraic analysis, introduced by I. Bernstein and M. Sato, see [18, 57]. The existence of a 'p-adic pseudodifferential analog' of Sato-Bernstein's D-module theory would have serious mathematical implications. We explain briefly this problem here. Let $f(x)$ be a polynomial with coefficients in a field of characteristic zero. I.N. Bernstein established the existence of an operator $P(s, x_1, \ldots, x_n, \frac{\partial}{\partial x_1}, \ldots, \frac{\partial}{\partial x_n}) := P$ such that

$$Pf(x)^{s+1} = b_f(s)f(x)^s, \tag{7.3}$$

where $b_f(s)$ is a polynomial, called the Sato-Bernstein polynomial of f, see [18], [57, Chapter 4]. In the Archimedean case, i.e. $\mathbb{R}$, $\mathbb{C}$, the meromorphic continuation of the local zeta function attached to f follows from (7.3), by integration by parts and the poles of the corresponding meromorphic continuation are controlled by the roots of the Sato-Bernstein polynomial of f, see e.g. [57, Theorems 5.3.1 and 5.3.2]. A very relevant and interesrting question is the following: Does a pseudodifferential analog of (7.3) exist over non-Archimedean local fields of arbitrary characteristic? We are asking for a pseudodifferential operator $P(\partial)$ such that $P(\partial)|f(x)|_p^{s+1} = b_f(s)|f(x)|_p^s$ as distributions on $\mathcal{D}(\mathbb{Q}_p^n)$. A crucial step is the existence of an adjoint operator for $P(\partial)$, i.e. $\left(P(\partial)|f(x)|_p^{s+1}, \varphi\right) = \left(|f(x)|_p^{s+1}, P^*(\partial)\varphi\right)$, this adjoint operator exists in the space $\mathcal{H}_{\mathbb{C}}(\infty)$ introduced in [117].

7.3   Parabolic-Type Equations and Markov Processes

From the point of view of the applications of non-Archimedean analysis to physics, the parabolic-type equations and their Markov processes play a central role. The theory presented in Chaps. 2 and 3 is far from being complete. To construct a general theory for this type of equations is a natural task. The author believes that a non-Archimedean theory similar to the one presented by N. Jacob in [60] is possible.

7.4   Non-Archimedean Quantum Fields

The quantization of solutions of the Klein-Gordon equation introduced in Chap. 6 is a natural task. It seems that the construction of a neutral (real) quantum scalar field with mass parameter $m \in \mathbb{Q}_p^\times$ can be carried out using the machinery of the second quantization starting with $\mathcal{H} = L^2\left(V_{m^2}^+, d\lambda_{m^2}\right)$, the state space for a single spin-zero particle of mass m, see e.g. [34, 44, 96]. In [117], we construct p-adic Euclidean random fields Φ over $\mathbb{Q}_p^n$, for arbitrary n, these fields are solutions of p -adic stochastic pseudodifferential equations. From a mathematical perspective, the Euclidean fields are generalized stochastic processes parametrized by functions belonging to a nuclear countably Hilbert space, these spaces are introduced in this article, in addition, the Euclidean fields are invariant under the action of certain group of transformations. All these results were obtained for massive fields, more precisely when the mass parameter m is a positive real number. A natural problem is to extend these constructions to the case $m = 0$.

References

1. Albeverio, S., Belopolskaya, Y.: Stochastic processes in $\mathbb{Q}_p$ associated with systems of nonlinear PIDEs. p-Adic Numbers Ultrametric Anal. Appl. **1**(2), 105–117 (2009)
2. Albeverio, S., Karwowski, W.: Diffusion in p-adic numbers. In: Ito, K., Hida, H. (eds.) Gaussian Random Fields, pp. 86–99. World Scientific, Singapore (1991)
3. Albeverio, S., Karwowski, W.: A random walk on p-adics: the generator and its spectrum. Stoch. Process. Appl. **53**, 1–22 (1994)
4. Albeverio, S., Karwowski, W.: Jump processes on leaves of multibranching trees. J. Math. Phys. **49**(9), 093503, 20 pp. (2008)
5. Albeverio, S., Khrennikov, A.Yu., Shelkovich, V.M.: Theory of p-Adic Distributions: Linear and Nonlinear Models. Cambridge University Press, Cambridge (2010)
6. Ansari, A., Berendzen, J., Bowne, S.F., Frauenfelder, H., Iben, I.E.T., Sauke, T.B., Shyamsunder, E., Young, R.D.: Protein states and proteinquakes. Proc. Natl. Acad. Sci. USA **82**, 5000–5004 (1985)
7. Arendt, W., Batty, C.J.K., Hieber, M., Neubrander, F.: Vector-Valued Laplace Transforms and Cauchy Problems. Birkhäuser-Springer, Basel (2011)
8. Avetisov, V.A., Bikulov, A.H., Kozyrev, S.V.: Application of p-adic analysis to models of breaking of replica symmetry. J. Phys. A **32**(50), 8785–8791 (1999)
9. Avetisov, V.A., Bikulov, A.Kh., Kozyrev, S.V.: Description of logarithmic relaxation by a model of a hierarchical random walk. Dokl. Akad. Nauk **368**(2), 164–167 (1999, in Russian)
10. Avetisov, V.A., Bikulov, A.H., Kozyrev, S.V., Osipov, V.A.: p-Adic models of ultrametric diffusion constrained by hierarchical energy landscapes. J. Phys. A **35**(2), 177–189 (2002)
11. Avetisov, V.A., Bikulov, A.Kh., Osipov, V.A.: p-Adic description of characteristic relaxation in complex systems. J. Phys. A **36**(15), 4239–4246 (2003)
12. Avetisov, V.A., Bikulov, A.Kh., Osipov, V.A.: p-Adic models of ultrametric diffusion in the conformational dynamics of macromolecules. Proc. Steklov Inst. Math. **2452**, 48–57 (2004)
13. Avetisov, V.A., Bikulov, A.Kh.: On the ultrametricity of the fluctuation dynamic mobility of protein molecules. Proc. Steklov Inst. Math. **265**(1), 75–81 (2009)
14. Avetisov, V.A., Bikulov, A.Kh., Zubarev, A.P.: First passage time distribution and the number of returns for ultrametric random walks. J. Phys. A **42**(8), 085003, 18 pp. (2009)
15. Atiyah, M.F.: Resolution of singularities and division of distributions. Commun. Pure Appl. Math. **23**,145–150 (1970)
16. Bass, R.F., Levin, D.A.: Transition probabilities for symmetric jump processes. Trans. Am. Math. Soc. **354**(7), 2933–2953 (2002)

© Springer International Publishing AG 2016

W.A. Zúñiga-Galindo, *Pseudodifferential Equations Over Non-Archimedean Spaces*, Lecture Notes in Mathematics 2174, DOI 10.1007/978-3-319-46738-2

171

17. Berg, C., Forst, G.: Potential Theory on Locally Compact Abelian Groups. Springer, Berlin (1975)
18. Bernstein, I.N.: Modules over the ring of differential operators; the study of fundamental solutions of equations with constant coefficients. Funct. Anal. Appl. **5**(2), 1–16 (1972)
19. Beloshapka, O.: Feynman formulas for an infinite-dimensional p-adic heat type equation. Infin. Dimens. Anal. Quantum Probab. Relat. Top. **14**(1), 137–148 (2011)
20. Blair, A.D.: Adelic path integrals. Rev. Math. Phys. **7**, 21–49 (1995)
21. Bikulov, A.Kh., Volovich, I.V.: p-Adic Brownian motion. Izv. Math. **61**(3), 537–552 (1997)
22. Borevich, Z.I., Shafarevich, I.R.: Number Theory. Academic, London (1986)
23. Casas-Sánchez, O.F., Zúñiga-Galindo, W.A.: p-Adic elliptic quadratic forms, parabolic-type pseudodifferential equations with variable coefficients and Markov processes. p-Adic Numbers Ultrametric Anal. Appl. **6**(1), 1–20 (2014)
24. Cazenave, T., Haraux, A.: An Introduction to Semilinear Evolution Equations. Oxford University Press, Oxford (1998)
25. Chacón-Cortes, L.F., Zúñiga-Galindo, W.A.: Nonlocal operators, parabolic-type equations, and ultrametric random walks. J. Math. Phys. **54**, 113503 (2013) [Erratum **55**, 109901 (2014)]
26. Chacón-Cortes, L., Zúñiga-Galindo, W.A.: Non-local operators, non-Archimedean parabolic-type equations with variable coefficients and Markov processes. Publ. Res. Inst. Math. Sci. **51**(2), 289–317 (2015)
27. Chacón-Cortes, L., Zúñiga-Galindo, W.A.: Heat traces and spectral zeta functions for p-adic laplacians. Accepted in Algebra i Analiz. arXiv:1511.02146
28. Chen, Z.-Q., Kumagai, T.: Heat kernel estimates for jump processes of mixed types on metric measure spaces. Probab. Theory Relat. Fields **140**(1–2), 277–317 (2008)
29. Connes, A.: Trace formula in noncommutative geometry and the zeros of the Riemann zeta function. Sel. Math. (N.S.) **5**, 29–106 (1999)
30. Denef, J.: Report on Igusa's local zeta function. Séminaire Bourbaki **43**, exp. 741 (1990–1991); Astérisque **201–202–203**, 359–386 (1991). Available at http://www.wis.kuleuven.ac.be/algebra/denef.html
31. de Jager, E.M.: The Lorentz-invariant solutions of the Klein-Gordon equation. SIAM J. Appl. Math. **15**, 944–963 (1967)
32. de Jager, E.M.: Applications of Distributions in Mathematical Physics. Mathematical Centre Tracts, vol. 10. Mathematisch Centrum, Amsterdam (1964)
33. Diamond, H.: Elementary methods in the study of the distribution of prime numbers. Bull. Am. Math. Soc. (N.S.) **7**(3), 553–589 (1982)
34. Dimock, J.: Quantum Mechanics and Quantum Field Theory: A Mathematical Primer. Cambridge University Press, Cambridge (2011)
35. Dragovich, B.: p-Adic and adelic quantum mechanics. Proc. Steklov Inst. Math. **245**(2), 64–77 (2004)
36. Dragovich, B., Khrennikov, A.Yu., Kozyrev, S.V., Volovich, I.V.: On p-adic mathematical physics. p-Adic Numbers Ultrametric Anal. Appl. **1**(1), 1–17 (2009)
37. Dragovich, B., Radyno, Y., Khrennikov, A.: Generalized functions on adeles. J. Math. Sci. (N.Y.) **142**(3), 2105–2112 (2007)
38. Droniou, J., Gallouet, T., Vovelle, J.: Global solution and smoothing effect for a non-local regularization of a hyperbolic equation. J. Evol. Equ. **3**(3), 499–521 (2003)
39. Dynkin, E.B.: Markov Processes, vol. I. Springer, Berlin (1965)
40. Ehrenpreis, L.: Solution of some problems of division. Part I. Division by a polynomial of derivation. Am. J. Math. **76**, 883–903 (1954)
41. Engel, K.-J., Nagel, R.: One-Parameter Semigroups for Linear Evolution Equations. Springer, New York (2000)
42. Evans, S.N.: Local properties of Lévy processes on a totally disconnected group. J. Theor. Probab. **2**(2), 209–259 (1989)
43. Evans, S.N.: Local field Brownian motion. J. Theor. Probab. **6**(4), 817–850 (1993)
44. Folland, G.B.: Quantum Field Theory: A Tourist Guide for Mathematicians. American Mathematical Society, Providence (2008)

45. Friedman, A.: Partial Differential Equations of Parabolic Type. Prentice-Hall, Englewood Cliffs (1964)
46. Frauenfelder, H., McMahon, B.H., Fenimore, P.W.: Myoglobin, the hydrogen atom of biology and paradigm of complexity. Proc. Natl. Acad. Sci. USA **100**(15), 8615–8617 (2003)
47. Frauenfelder, H., Sligar, S.G., Wolynes, P.G.: The energy landscape and motions of proteins. Science **254**, 1598–1603 (1991)
48. Galeano-Peñaloza, J., Zúñiga-Galindo, W.A.: Pseudo-differential operators with semi-quasielliptic symbols over p-adic fields: J. Math. Anal. Appl. **386**(1), 32–49 (2012)
49. Gel'fand, I.M., Shilov, G.E.: Generalized Functions, vol. 1. Academic, New York (1977)
50. Goldfeld, D., Hundley, J.: Automorphic Representations and L-Functions for the General Linear Group, vol. I. Cambridge University Press, Cambridge (2011)
51. Haran, S.: Potentials and explicit sums in arithmetic. Invent. Math. **101**, 797–703 (1990)
52. Haran, S.: Quantizations and symbolic calculus over the p-adic numbers. Ann. Inst. Fourier **43**(4), 997–1053 (1993)
53. Harlow, D., Shenker, S., Stanford, D., Susskind, L.: Tree-like structure of eternal inflation: a solvable model. Phys. Rev. D **85**(6) (2012). Article Number: 063516
54. Hironaka, H.: Resolution of singularities of an algebraic variety over a field of characteristic zero. Ann. Math. **79**, 109–326 (1964)
55. Hörmander, L.: On the division of distributions by polynomials. Ark. Mat. **3**, 555–568 (1958)
56. Hörmander, L.: The Analysis of Linear Partial Differential Operators. II: Differential Operators with Constant Coefficients. Grundlehren der Mathematischen Wissenschaften, vol. 257. Springer, Berlin (1983)
57. Igusa, J.-I.: An Introduction to the Theory of Local Zeta Functions. AMS/IP Studies in Advanced Mathematics, vol. 14. American Mathematical Society, Providence (2000)
58. Igusa, J.-I.: Some aspects of the arithmetic theory of polynomials. Discrete Groups in Geometry and Analysis (New Haven, 1984). Progress in Mathematics, vol. 67, pp. 20–47. Birkhäuser, Boston (1987)
59. Igusa, J.-I.: Zeta distributions associated with some invariants. Am. J. Math. **109**(1), 1–33 (1987)
60. Jacob, N.: Pseudo Differential Operators and Markov Processes, Vol. II: Generators and Their Potential Theory, xxii+453 pp. Imperial College Press, London (2002)
61. Karwowski, W.: Diffusion processes with ultrametric jumps. Rep. Math. Phys. **60**(2), 221–235 (2007)
62. Karwowski, W., Mendes, R.V.: Hierarchical structures and asymmetric stochastic processes on p-adics and adèles. J. Math. Phys. **35**(9), 4637–4650 (1994)
63. Kigami, J.: Transitions on a noncompact Cantor set and random walks on its defining tree. Ann. Inst. Henri Poincaré Probab. Stat. **49**(4), 1090–1129 (2013)
64. Khrennikov, A.: p-Adic Valued Distributions in Mathematical Physics. Kluwer, Dordrecht (1994)
65. Khrennikov, A.: Non-Archimedean Analysis: Quantum Paradoxes, Dynamical Systems and Biological Models. Kluwer, Dordrecht (1997)
66. Khrennikov, A.Yu., Kozyrev, S.V.: Wavelets on ultrametric spaces. Appl. Comput. Harmon. Anal. **19**, 61–76 (2005)
67. Khrennikov, A.Yu., Kozyrev, S.V.: Replica symmetry breaking related to a general ultrametric space I: replica matrices and functionals. Phys. A: Stat. Mech. Appl. **359**, 222–240 (2006)
68. Khrennikov, A.Yu., Kozyrev, S.V.: Replica symmetry breaking related to a general ultrametric space II: RSB solutions and the $n \to 0$ limit. Phys. A: Stat. Mech. Appl. **359**, 241–266 (2006)
69. Khrennikov, A.Yu., Kozyrev, S.V.: Replica symmetry breaking related to a general ultrametric space III: the case of general measure. Phys. A: Stat. Mech. Appl. **378**(2), 283–298 (2007)
70. Khrennikov, A.Yu., Kozyrev, S.V.: Ultrametric random field. Infin. Dimens. Anal. Quantum Probab. Relat. Top. **9**(2), 199–213 (2006)

71. Khrennikov, A.Yu., Kozyrev, S.V., Oleschko, K., Jaramillo, A.G., Correa, L.J.: Application of p-adic analysis to time series. Infin. Dimens. Anal. Quantum Probab. Relat. Top. **16**(4), 1350030, 15 pp. (2013)
72. Khrennikov, A.Y., Radyno, Y.V.: On adelic analogue of Laplacian. Proc. Jangjeon Math. Soc. **6**(1), 1–18 (2003)
73. Khrennikov, A.Y., Shelkovich, V.M., van der Walt, J.H.: Adelic multiresolution analysis, construction of wavelet bases and pseudo-differential operators. J. Fourier Anal. Appl. **19**(6), 1323–1358 (2013)
74. Khrennikov, A.Yu., Kosyak, A.V., Shelkovich, V.M.: Wavelet analysis on adeles and pseudo-differential operators. J. Fourier Anal. Appl. **18**(6), 1215–1264 (2012)
75. Khrennikov, A.Yu., Kozyrev, S.V.: Genetic code on the diadic plane. Phys. A: Stat. Mech. Appl. **381**, 265–272 (2007)
76. Khrennikov, A.Yu., Kozyrev, S.V.: 2-Adic clustering of the PAM matrix. J. Theor. Biol. **261**, 396–406 (2009)
77. Kozyrev, S.V., Khrennikov, A.Yu.: Pseudodifferential operators on ultrametric spaces, and ultrametric wavelets. Izv. Math. **69**(5), 989–1003 (2005)
78. Khrennikov, A.Yu., Kozyrev, S.V.: p-Adic pseudodifferential operators and analytic continuation of replica matrices. Theor. Math. Phys. **144**(2), 1166–1170 (2005)
79. Kochubei, A.N.: A non-Archimedean wave equation. Pac. J. Math. **235**, 245–261 (2008)
80. Kochubei, A.N.: Pseudo-Differential Equations and Stochastics over Non-Archimedean Fields. Marcel Dekker, New York (2001)
81. Kochubei, A.N., Parabolic equations over the field of p-adic numbers. Math. USSR Izv. **39**, 1263–1280 (1992)
82. Kozyrev, S.V.: Methods and applications of ultrametric and p-adic analysis: from wavelet theory to biophysics. Proc. Steklov Inst. Math. **274**(1 Suppl.), 1–84 (2011)
83. Leichtnam, E.: Scaling group flow and Lefschetz trace formula for laminated spaces with p-adic transversal. Bull. Sci. Math. **131**(7), 638–669 (2007)
84. Malgrange, B.: Existence et approximation des solutions des é quations aux dérivées partielles et des équations de convolution. Ann. Inst. Fourier **6**, 271–355 (1955/1956)
85. Manin, Y.I.: Reflections on Arithmetical Physics. Conformal Invariance and String Theory, pp. 293–303. Academic, New York (1989)
86. Mézard, M., Parisi, G.: Virasoro Miguel Angel. Spin Glass Theory and Beyond. World Scientific, Singapore (1987)
87. Ogielski, A.T., Stein, D.L.: Dynamics on ultrametric spaces. Phys. Rev. Lett. **55**(15), 1634–1637 (1985)
88. Ono, T.: Gauss transforms and zeta-functions. Ann. Math. **91**(2), 332–361 (1970)
89. Ortner, N., Wagner, P.: A short proof of the Malgrange-Ehrenpreis theorem. In: Dierolf, S., Dineen, S., Domański, P. (eds.) Functional Analysis. Proceedings of the 1st International Workshop in Trier, Germany, 1994, pp. 343–352. de Gruyter, Berlin (1996)
90. Parisi, G., Sourlas N.: p-Adic numbers and replica symmetry breaking. Eur. Phys. J. B Condens. Matter Phys. **14**(3), 535–542 (2000)
91. Radyno, Y.V., Radyna, Y.M.: Generalized Functions on Adeles. Linear and Non-linear Theories. Linear and Non-linear Theory of Generalized Functions and Its Applications. Banach Center Publications, vol. 88, pp. 243–250. Polish Academy of Sciences, Institute of Mathematics, Warsaw (2010)
92. Rallis, S., Schiffmann, G.: Distributions invariantes par le groupe orthogonal. Analyse Harmonique Sur Les Groupes de Lie (Sém., Nancy-Strasbourg, 1973–1975). Lecture Notes in Mathematics, vol. 497, pp. 494–642. Springer, New York (1975)
93. Ramakrishnan, D., Valenza, R.J.: Fourier Analysis on Number Fields. Springer, New York (1999)
94. Rammal, R., Toulouse, G., Virasoro, M.A.: Ultrametricity for physicists. Rev. Mod. Phys. **58**(3), 765–788 (1986)
95. Reed, M., Simon, B.: Methods of Modern Mathematical Physics: Functional Analysis I. Academic, New York (1980)

96. Reed, M., Simon, B.: Methods of Modern Mathematical Physics, Vol. II: Fourier Analysis, Self-Adjointness. Academic, New York (1975)

97. Rodríguez-Vega, J.J.: On a general type of p-adic parabolic equations. Rev. Colomb. Mat. **43**(2), 101–114 (2009)

98. Rodríguez-Vega, J.J., Zúñiga-Galindo, W.A.: Taibleson operators, p-adic parabolic equations and ultrametric diffusion. Pac. J. Math. **237**(2), 327–347 (2008)

99. Rosay, J.-P.: A very elementary proof of the Malgrange-Ehrenpreis theorem. Am. Math. Mon. **98**(6), 518–523 (1991)

100. Rudin, W.: Fourier Analysis on Groups. Interscience, New York (1962)

101. Samko, S.G.: Hypersingular Integrals and Their Applications. Taylor and Francis, London (2002)

102. Samko, S.G., Kilbas, A.A., Marichev, O.I.: Fractional Integrals and Derivatives and Some of Their Applications. Gordon and Breach Science Publishers, Yverdon (1993)

103. Schweber, S.S.: An Introduction to Relativistic Quantum Field Theory. Row/Peterson, Evanston (1961)

104. Serre, J.-P.: Lie Algebras and Lie Groups. American Mathematical Society, Providence (1968)

105. Taibleson, M.H.: Fourier Analysis on Local Fields. Princeton University Press, Princeton (1975)

106. Torba, S.M., Zúñiga-Galindo, W.A.: Parabolic type equations and Markov stochastic processes on adeles. J. Fourier Anal. Appl. **19**(4), 792–835 (2013)

107. Varadarajan, V.S.: Reflections on Quanta, Symmetries, and Supersymmetries. Springer, New York (2011)

108. Varadarajan, V.S.: Path integrals for a class of p-adic Schr ödinger equations. Lett. Math. Phys. **39**(2), 97–106 (1997)

109. Varadarajan, V.S.: Arithmetic quantum physics: why, what, and whither. Proc. Steklov Inst. Math. **245**(2), 258–265 (2004)

110. Veys, W., Zúñiga-Galindo, W.A.: Zeta functions for analytic mappings, log-principalization of ideals, and newton polyhedra. Trans. Am. Math. Soc. **360**, 2205–2227 (2008)

111. Vladimirov, V.S., Volovich, I.V., Zelenov, E.I.: p-Adic Analysis and Mathematical Physics. World Scientific, Singapore (1994)

112. Volovich, I.V.: Number theory as the ultimate physical theory. p-adic Numbers Ultrametric Anal. Appl. **2**(1), 77–87 (2010)

113. Volovich, I.V.: p-Adic string. Class. Quantum Grav. **4**, L83–L87 (1987)

114. Wales, D.J., Miller, M.A., Walsh, T.R.: Archetypal energy landscapes. Nature **394**, 758–760 (1998)

115. Weil, A.: Basic Number Theory. Springer, New York (1967)

116. Yasuda, K.: Markov processes on the adeles and representations of Euler products. J. Theor. Probab. **23**(3), 748–769 (2010)

117. Zúñiga-Galindo, W.A.: Non-Archimedean white noise, pseudodifferential stochastic equations, and massive Euclidean fields. J. Fourier Anal. Appl. (2016). doi:10.1007/s00041-016-9470-1

118. Zúñiga-Galindo, W.A.: The non-Archimedean stochastic heat equation driven by Gaussian noise. J. Fourier Anal. Appl. **21**(3), 600–627 (2015)

119. Zúñiga-Galindo, W.A.: The Cauchy problem for non-Archimedean pseudodifferential equations of Klein-Gordon type. J. Math. Anal. Appl. **420**(2), 1033–1050 (2014)

120. Zúñiga-Galindo, W.A.: Local zeta functions and fundamental solutions for pseudo-differential operators over p-adic fields. p-Adic Numbers Ultrametric Anal. Appl. **3**(4), 344–358 (2011)

121. Zúñiga-Galindo, W.A.: Local zeta functions supported on analytic sets and Newton polyhedra. Int. Math. Res. Not. IMRN (15), 2855–2898 (2009)

122. Zúñiga-Galindo, W.A.: Parabolic equations and Markov processes over p-adic fields. Potential Anal. **28**(2), 185–200 (2008)

123. Zúñiga-Galindo, W.A.: Fundamental solutions of pseudo-differential operators over p-adic fields. Rend. Sem. Mat. Univ. Padova **109**, 241–245 (2003)

LECTURE NOTES IN MATHEMATICS ♘ Springer

Editors in Chief: J.-M. Morel, B. Teissier;

Editorial Policy

1. Lecture Notes aim to report new developments in all areas of mathematics and their applications – quickly, informally and at a high level. Mathematical texts analysing new developments in modelling and numerical simulation are welcome.

   Manuscripts should be reasonably self-contained and rounded off. Thus they may, and often will, present not only results of the author but also related work by other people. They may be based on specialised lecture courses. Furthermore, the manuscripts should provide sufficient motivation, examples and applications. This clearly distinguishes Lecture Notes from journal articles or technical reports which normally are very concise. Articles intended for a journal but too long to be accepted by most journals, usually do not have this "lecture notes" character. For similar reasons it is unusual for doctoral theses to be accepted for the Lecture Notes series, though habilitation theses may be appropriate.

2. Besides monographs, multi-author manuscripts resulting from SUMMER SCHOOLS or similar INTENSIVE COURSES are welcome, provided their objective was held to present an active mathematical topic to an audience at the beginning or intermediate graduate level (a list of participants should be provided).

   The resulting manuscript should not be just a collection of course notes, but should require advance planning and coordination among the main lecturers. The subject matter should dictate the structure of the book. This structure should be motivated and explained in a scientific introduction, and the notation, references, index and formulation of results should be, if possible, unified by the editors. Each contribution should have an abstract and an introduction referring to the other contributions. In other words, more preparatory work must go into a multi-authored volume than simply assembling a disparate collection of papers, communicated at the event.

3. Manuscripts should be submitted either online at www.editorialmanager.com/lnm to Springer's mathematics editorial in Heidelberg, or electronically to one of the series editors. Authors should be aware that incomplete or insufficiently close-to-final manuscripts almost always result in longer refereeing times and nevertheless unclear referees' recommendations, making further refereeing of a final draft necessary. The strict minimum amount of material that will be considered should include a detailed outline describing the planned contents of each chapter, a bibliography and several sample chapters. Parallel submission of a manuscript to another publisher while under consideration for LNM is not acceptable and can lead to rejection.

4. In general, **monographs** will be sent out to at least 2 external referees for evaluation.

   A final decision to publish can be made only on the basis of the complete manuscript, however a refereeing process leading to a preliminary decision can be based on a pre-final or incomplete manuscript.

   Volume Editors of **multi-author works** are expected to arrange for the refereeing, to the usual scientific standards, of the individual contributions. If the resulting reports can be forwarded to the LNM Editorial Board, this is very helpful. If no reports are forwarded or if other questions remain unclear in respect of homogeneity etc, the series editors may wish to consult external referees for an overall evaluation of the volume.

5. Manuscripts should in general be submitted in English. Final manuscripts should contain at least 100 pages of mathematical text and should always include

    – a table of contents;
    – an informative introduction, with adequate motivation and perhaps some historical remarks: it should be accessible to a reader not intimately familiar with the topic treated;
    – a subject index: as a rule this is genuinely helpful for the reader.
    – For evaluation purposes, manuscripts should be submitted as pdf files.

6. Careful preparation of the manuscripts will help keep production time short besides ensuring satisfactory appearance of the finished book in print and online. After acceptance of the manuscript authors will be asked to prepare the final LaTeX source files (see LaTeX templates online: https://www.springer.com/gb/authors-editors/book-authors-editors/manuscriptpreparation/5636) plus the corresponding pdf- or zipped ps-file. The LaTeX source files are essential for producing the full-text online version of the book, see http://link.springer.com/bookseries/304 for the existing online volumes of LNM). The technical production of a Lecture Notes volume takes approximately 12 weeks. Additional instructions, if necessary, are available on request from lnm@springer.com.

7. Authors receive a total of 30 free copies of their volume and free access to their book on SpringerLink, but no royalties. They are entitled to a discount of 33.3 % on the price of Springer books purchased for their personal use, if ordering directly from Springer.

8. Commitment to publish is made by a *Publishing Agreement*; contributing authors of multiauthor books are requested to sign a *Consent to Publish form*. Springer-Verlag registers the copyright for each volume. Authors are free to reuse material contained in their LNM volumes in later publications: a brief written (or e-mail) request for formal permission is sufficient.

Addresses:
Professor Jean-Michel Morel, CMLA, École Normale Supérieure de Cachan, France
E-mail: moreljeanmichel@gmail.com

Professor Bernard Teissier, Equipe Géométrie et Dynamique,
Institut de Mathématiques de Jussieu – Paris Rive Gauche, Paris, France
E-mail: bernard.teissier@imj-prg.fr

Springer: Ute McCrory, Mathematics, Heidelberg, Germany,
E-mail: lnm@springer.com

Printed in the United States
By Bookmasters